AF323851

Noise Contr
in Buildings

Fundamental and Applica

Noise Control in Buildings

Fundamental and Applications

Mahavir Singh

Narosa Publishing House

New Delhi   Chennai   Mumbai   Kolkata

Noise Control in Buildings
Fundamental and Applications
172 pgs. | 90 figs. | 26 tbls.

Mahavir Singh
Acoustics, Ultrasonics and Vibration Section
CSIR-National Physical Laboratory
New Delhi

N A R O S A P U B L I S H I N G H O U S E P V T. LT D.

22 Delhi Medical Association Road, Daryaganj, New Delhi 110 002
35-36 Greams Road, Thousand Lights, Chennai 600 006
306 Shiv Centre, Sector 17, Vashi, Navi Mumbai 400 703
2F-2G Shivam Chambers, 53 Syed Amir Ali Avenue, Kolkata 700 019

www.narosa.com

ISBN 978-81-8487-342-9

Published by N.K. Mehra for Narosa Publishing House Pvt. Ltd.,
22 Delhi Medical Association Road, Daryaganj, New Delhi 110 002

Printed in India

ACKNOWLEDGEMENTS

The author would like to acknowledge the support of the Acoustics Section, and particularly Dr. V. Mohanan, for their support over the years in the measurements of sound transmission through buildings, and for giving permission for extracts and results from unpublished technical reports to be included in this book.

The author would also like to acknowledge the support of the BPB India Gypsum and Lafarge Boral Gypsum India, for giving permission to include results from an unpublished technical report in Chapters 3 and 4.

Much of the basic work on the measurements of sound transmission in buildings has been funded by a series of research grants from BPB India Gypsum over a period of 18 years.

The author is grateful for the assistance of Dr. Omkar Sharma, who carried out much of the work for the Acoustics, Ultrasonics & Vibration Section under a series of research contracts, to Dr. Kirti Soni, who also supplied data and read through the manuscript and to Dr. Yudhisther Kumar who checked the final manuscript. Recent research data was also given by my research student, Mr. D. P. Singh, and Surjit Singh which enhanced the sections in Chapter 1 to 7.

Special thanks are also expressed to Prof. K. K. Pujara, IIT Delhi for giving fruitful suggestions in the review and correction of the entire text.

Mahavir Singh

PREFACE

Noise Control in Buildings contains a complete set of data on the properties of acoustical materials and on the sound insulation of walls and floor/ceiling constructions. This wealth of technical information provides an invaluable resource for the professional as well as the non-professional:

- The properties and selection of acoustical materials
- The design of select wall and floor/ceiling constructions
- Airborne sound insulation
- Control of noise communicated by building structures
- Acoustical characteristics of rooms

Also included are effective methods for dealing with noise problems in HVAC systems, plumbing systems, and machinery - plus innovations in techniques for the design of buildings with low noise levels.

Controlling noise in buildings is an important part of an architect's and engineer's responsibility. This book provides practical guidelines on avoiding noise problems during the design and construction of new buildings, and eliminating noise in existing structures. Noise control in buildings covers such topics as properties of sound absorptive materials, acoustical characteristics of rooms, airborne sound insulation, and structure-borne sound insulation. Included are proven methods- and technical data-for dealing with noise, HVAC systems, plumbing systems, and machinery, plus information on the design of buildings for noise control.

It provides a wealth of information, mainly without the use of mathematics that can be used to ensure that noise control measures are incorporated into buildings during the design stage. The reader will find introductory material on the general aspects of sound transmission into buildings, properties of sound waves, sound absorption coefficients, and tables of sound absorption coefficients.

It is pointed out in the preface that extensive tables of the sound absorptive properties of materials are not generally available in the literature, and that the characteristics of acoustical materials manufactured have been collected and published in Chapters 2, 3 and 4. Sound in Room, Developments in Noise Control and Sound Transmission through Building Elements respectively.

Mahavir Singh

CONTENTS

1

BASICS OF NOISE CONTROL

1.1 INTRODUCTION

Sound is a pervasive phenomenon and such factors as the high density of new residential construction and the use of high powered equipment in modern lightweight construction systems are leading to higher noise levels in our homes, our work spaces and generally everywhere in our cities. In fact, surveys done in the United States report that the noise level of major cities is rising at the rate of approximately one decibel per year. We can no longer ignore the trend.

While the general field of noise control is quite broad, the Insight program was limited to noise control in buildings; however, the principles of noise control are the same regardless of the environment being considered.

Inadequate noise control has often made a restaurant meal an unpleasant experience. High noise levels in the workplace are distracting and irritating, possibly resulting in higher absenteeism and loss of productivity. Inadequate acoustical privacy is common in office buildings. Conference rooms, despite the fact that their primary purpose is speech communication, are often excessively reverberant and noisy, making it difficult to understand speech from people only a short distance away. Most schools, churches, gymnasiums and swimming pools, as they now exist, could benefit from an acoustical tune-up, improving the quality of their indoor environment and their revenue-earning potential.

A recent survey in Vancouver found about 40% of the tenants in several multi-family dwelling projects dissatisfied with the degree of acoustical privacy in their homes. This was the major problem as far as the tenants were concerned, and it is a major complaint in many of our new apartment and condominium projects.

Elsewhere, in a large hotel in India, the acoustical engineers were overruled by the project designers; in order to save some ₹ 3.7 crores they provided bedrooms with single glazing and operable sashes for ventilation. When the hotel came into service, most of the rooms had so much traffic noise penetrating, that they were unusable for sleeping. To rectify the situation, the hotel had to install double glazing and air conditioning at a cost of ₹ 9.6 crores, not including the loss of revenue and general disturbance during the repair work.

Such errors are inexcusable, given that the knowledge exists to design and build quality acoustical environments.

The essence of this Insight program is a discussion of the principles of noise control in buildings. The fundamentals of sound are explained in the first paper, followed by the criteria and methods to control noise within rooms, in the second paper. The third paper examines the issue of noise reduction across walls, floors, windows and doors, and the last paper deals with acoustics of buildings, with an emphasis on machine and plumbing noises and the annoying problem of flanking noise.

1.2 PHYSICAL PROPERTIES OF SOUND

The material in this section gives a brief introduction to some of the terms used in building acoustics and the test methods used to characterize systems and materials. A very basic understanding of some fundamentals of acoustics and terms used in building acoustics, is all that is necessary to understand the material in this and the following chapters. To emphasize the simplicity of the approach, equations are kept to a minimum.

Sound is generated by creating a disturbance of the air, which sets up a series of pressure waves fluctuating above and below the air's normal atmospheric pressure, much as a stone that falls in water generates expanding ripples on the surface. Unlike the water waves, however, these pressure waves propagate in all directions from the source of the sound. Our ears sense these pressure fluctuations, convert them to electrical impulses, and send them to our brain, where they are interpreted as sound.

There are many sources of sound in buildings: voices, human activities, external noises such as traffic, entertainment devices and machinery. They all generate small rapid variations in pressure about the static atmospheric pressure; these propagate through the air as sound waves.

As well as travelling in air, sound can travel as vibrational waves in solids or liquids. The terms airborne and structure-borne sound are used depending on which medium the sound is travelling in at the time. For example, the noise from a radio set may begin as airborne sound, enter the structure of the building and travel for some distance as structure-borne sound, and then be radiated again as airborne sound in another place. The importance of structure-borne sound will become more apparent when flanking sound transmission is shown in Figure 1.1.

Air pressure is usually measured in units of Pascals (Pa). Atmospheric pressure is about 100 kPa. Sound pressure is a measure of the fluctuation of the air pressure above and below normal atmospheric pressure as the sound waves propagate past a listener.

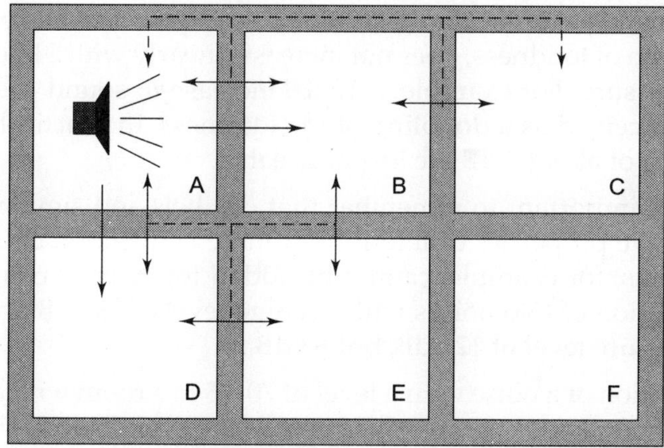

AIRBORNE AND STRUCTURE-BORNE SOUND

Figure 1.1

The pressure variations in an individual sound wave are much less than the static atmospheric pressure, but the range of sound pressures encountered in acoustics is very large. The threshold of hearing is assumed to correspond to pressure fluctuations of 20 Micro Pascals; some individuals will have more acute hearing than this, some less. The threshold of pain in the ear corresponds to pressure fluctuations of about 200 Pa. This second value is ten million times the first. These unwieldy numbers are converted to more convenient ones using a logarithmic scale, the decibel scale. Sound pressure levels are expressed as a number followed by the symbol dB. Sound level meters convert electrical signals from a microphone to sound pressure levels in dB. Table 1.1 gives some representative sound pressure levels encountered in a range of situations.

Table 1.1: Typical Sound Levels

Jet takeoff, artillery fire, riveting	120 or more
Rock band or very loud orchestra	100 – 120
Unruffled truck, police whistle	80 – 100
Average radio or TV	70 – 90
Human voice at 1 m	55 – 60
Background in private office	35 – 40
Quiet home	25 – 35
Threshold of hearing	20

Decibels are more easily related to the response of the human ear, which also responds logarithmically to sound. The response of our ears, that is, our perception of loudness, does not increase linearly with a linear increase in sound pressure. For example, a 10 dB increase in sound pressure level would be perceived as a doubling of the loudness. In practical situations, level changes of about 3 dB are just noticeable.

It is very important to remember that decibels and similar acoustical quantities have properties different from more conventional units. Sound pressure levels, for example, cannot be added together as can kilograms. The combination of two noises with average levels of 60 dB does not give a sound pressure level of 120 dB, but 63 dB.

The addition of a noise with a level of 70 dB to a room with a level of 80 dB will result in no measureable difference in the overall level. This does not mean, however, that a large number of secondary sources can be introduced into an environment without increasing the overall level. If ten 'negligible' 70 dB noise sources are combined with one 80 dB noise source, the resulting level will be 83 dB. Fortunately in building acoustics, there is seldom a need to combine noise levels in this way or to do many complicated calculations with decibels or other logarithmic units. These examples merely emphasize the peculiarities of the decibel scale.

1.3 FILTERS AND SPECTRA

The pressure variations associated with many sounds have a repeating pattern. This is especially true of musical sounds. The frequency of a repetitive or periodic phenomenon is the number of times per second chat the pattern repeats itself. High frequency or high-pitched sounds are heard when the pressure variations in the air occur rapidly-several thousand times per second. The unit used for frequency is hertz (Hz). Thus a musical note with a frequency of 1200 Hz has a fundamental pattern of pressure variations that repeats itself 1200 times per second. In building acoustics, it is important to know the frequencies which make up a sound because different frequencies behave differently.

Typical sounds in buildings are complex without any clearly defined pattern, and are usually classified as noise. There is usually no obvious frequency or tone associated with such noises, although there may be tonal characteristics such as 'rumbling' or 'hissing'. Despite this, it is possible to analyse noises in terms of the energy contained in certain frequency intervals. This is called spectral analysis.

A narrow band spectrum shows the energy present at each frequency in great detail. Figure 1.2(a) shows the repeating time pattern of a sound source and its corresponding narrow band spectrum. This kind of information is very useful income instances but more of ten the detail is overwhelming; in building acoustics it is common to present spectra in standard one-third-octave band or even octave band plots. The term octave has the same meaning in building acoustics as it does in music; adjacent octave bands have frequencies that differ by a factor of two. Adjacent one-third-octave bands are one-third of an octave apart To produce this kind of spectrum the energy in specified regions of the narrow band spectrum is added together to give the octave or third-octave level l. This type of presentation gives information that is easier to relate to the response of the ear. Figures 1.2(b) and (c) show the third-octave and octave band spectra obtained from the narrow-band spectrum in Figure 1.2(a).

In Figure 1.2, the horizontal axis represents the particular frequency or the frequency band of the noise. The vertical axis represents the sound pressure level or energy at the given frequency or frequency band.

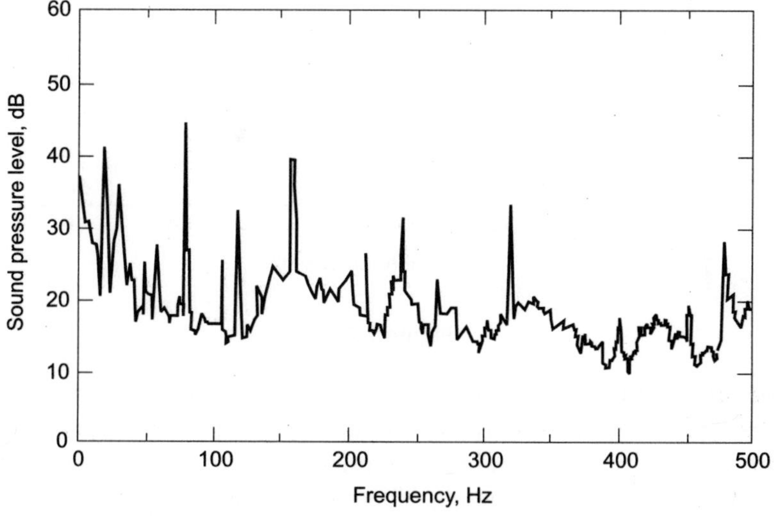

Figure 1.2(a)

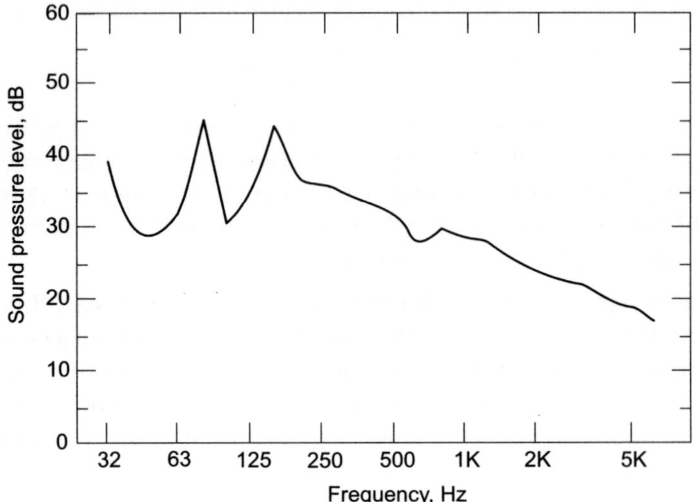

Figure 1.2(b)

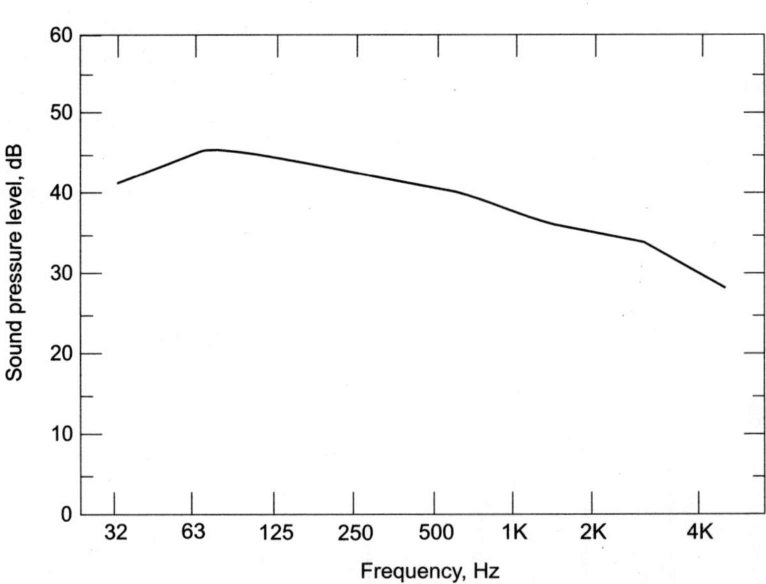

Figure 1.2(c)

The one-third-octave band spectrum for a particular noise source provides a good deal of information but a single number rating is often more convenient. One way of obtaining a single number describing a complex noise is to use A-weighted levels. The human ear is not equally sensitive at all frequencies; sounds of the same level but with different frequencies will not be considered equally loud. A sound at 3 kHz at a level of 54 dB, for example, will sound as loud as one at 50 Hz at a level of 79 dB.

The dashed curve in Figure 1.3 shows the idealised response of the human ear at intermediate sound levels. The solid curve approximates the response of the ear and is called the A-weighting curve.

The output of a microphone or sound level meter can be altered using an A-weighting electrical filter so that it more closely represents the response of the human ear. The resulting sound pressure levels are expressed as a number followed by the symbol dBA.

Noises with identical A-weighted levels can have quite different spectra and can evoke quite different responses from people. The use of such pejorative terms as whine, rasp, grate, rumble and hiss to describe sounds shows that people are well aware of the differences in spectral content. Other terms in common use such as loudness, noisiness and annoyance are not at all synonymous, nor clearly related to sound pressure levels. Despite the shortcomings of this simplistic method of rating noises, it is in common use.

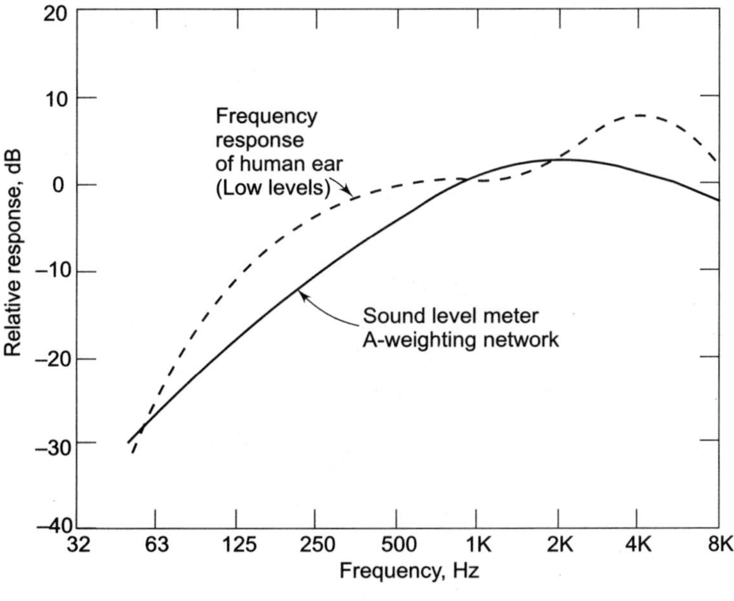

Figure 1.3

1.4   NOISE CRITERIA CONTOURS

Another method of describing the noise in buildings uses a set of octave band contours known as Noise Criteria or NC contours. In India, these are commonly used in Heating, Ventilating and Air Conditioning (HVAC) work and in other areas of building acoustics to specify maximum sound pressure levels in rooms. They provide a convenient way of rating the noise level, and to some extent, the spectrum in a room. Figure 1.4 shows two examples of how they are used. The spectrum for the sound of interest is drawn on the same graph paper as the NC contours. The point where the spectrum touches the highest NC curve, determines the NC rating for the spectrum. Estimates are made when the spectrum falls between the NC contours shown. Two spectra can have the same NC value but quite different shapes. Although the relationship between the A-weighted level and the NC number will depend on the actual spectrum, usually the A-weighted level is about 7 dB greater than the NC value. NC contours do not describe the spectrum of a neutral sounding noise to be sought after as a design goal. In fact, they have been described as' both rumbly and hissy'. Background noise spectra with less low and high frequency energy are usually found to be less objectionable.

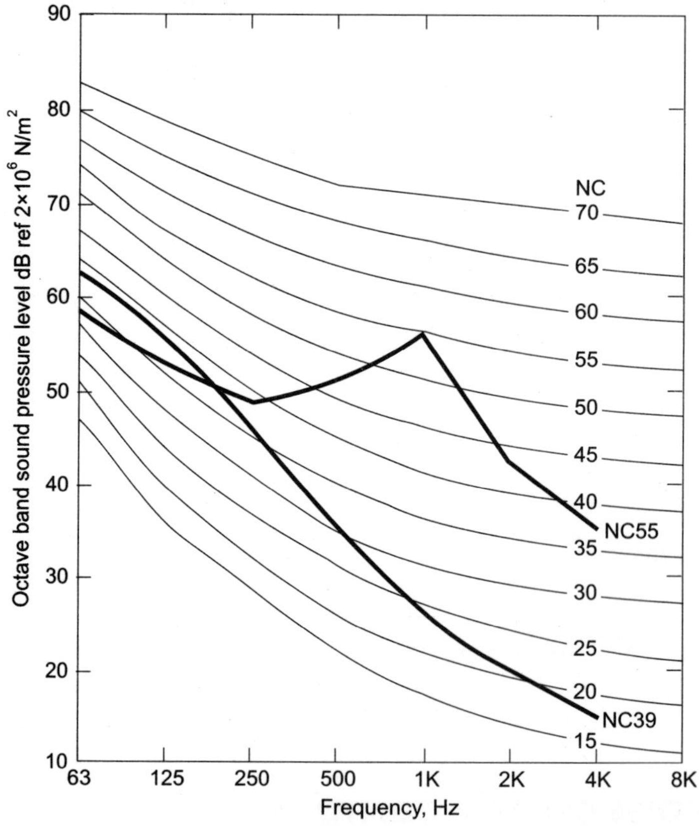

Figure 1.4

Recent issues of the ASHRAE guide include RC contours, which are intended to represent well-balanced, neutral sounding spectra. These ratings, meant to replace the NC contours and to give some indication of the acceptability of the background sound, are beginning to be used more widely in India. Despite this new system, more research needs to be done so that the subjective acceptability of noises in buildings can be rated more accurately.

1.5 STANDARD MEASUREMENTS FOR BUILDING ACOUSTICS

Acoustical design for buildings requires quantitative information about acoustical products, materials and systems so that recommended design criteria can be met. A number of acoustical test procedures are used in the laboratory and in buildings to obtain this kind of information and to verify that the building is performing as the designer intended.

For most noise control work in buildings, the two most important acoustical properties of the materials and systems used are sound-absorption and sound transmission loss. For machines to be used in buildings the important characteristic is the sound power.

Frequency Range

Measurements in building acoustics are usually made in one-third-octave bands from about 100 Hz to 4000 Hz. Although frequencies lower than 100 Hz occur in buildings, very low frequencies are not easily measured in any meaningful way because of the resonance effects in rooms. The upper frequency limit can go as high as 15000 Hz, when specialized measurements such as those dealing with concert halls are being made; frequencies down to the 31 Hz octave band are sometimes measured for HVAC work.

Sound Absorption

The sound-absorption coefficient is the fraction of incident sound energy that is absorbed by a material, sometimes expressed as a percentage. Sound is absorbed when part of the sound energy striking a surface or an object is converted into heat energy in the pores of the material. The absorption coefficient depends on the sound frequency and on the angle of incidence of the sound waves on the material. Absorption coefficients provided in manufacturers' literature is usually obtained by testing in a reverberation room (is shown Figure 1.5). This provides an average value for each one-third-octave band and all angles of incidence. Data are usually given at six standard frequencies (125, 250, 500, 1000, 2000 and 4000 Hz), although more modern laboratories will provide data at all one-third-octave bands from about 100 Hz to 5000 Hz. Figure 1.6 shows an example of how the results of an absorption test are plotted.

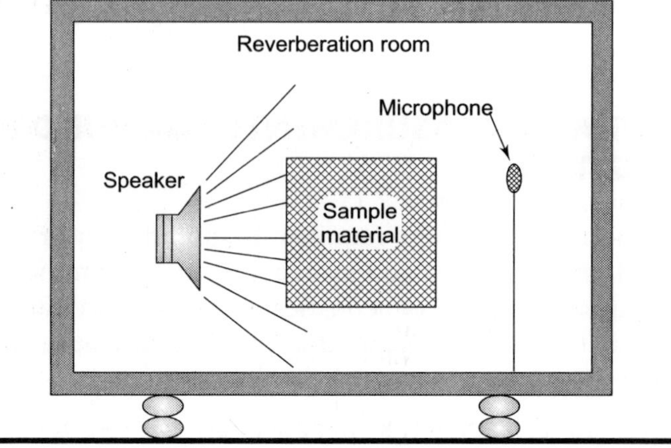

Figure 1.5

The sound absorption for a sample of material or an object is measured in sabins or metric sabins. One sabin maybe thought of as the absorption of unit area (1 m^2 or 1 ft^2) of a surface that has an absorption coefficient of 1.0 (100%). When areas are measured in square meters, the term metric sabin is used. The absorption for a surface can be found by multiplying its area by its absorption coefficient. Thus for a material with an absorption coefficient of 0.5, 10 ft^2 of this material has a sound absorption of 5 sabins and 100 m^2, of 50 metric sabins.

Noise Reduction Coefficient

The noise reduction coefficient (NRC), also shown in Figure 1.6, is the average of the four sound-absorption coefficients at 250, 500, 1000 and 2000 Hz. This single number rating is convenient for ranking, the average effectiveness of different materials. However, for a more complete acoustical design, it is necessary to consider individual coefficients at each frequency.

Sound Transmission Loss

When sound waves strike one side of a partition, the pressure variations cause vibrations in the partition, and part of the power in the sound wave is transferred to the partition. All or part of this vibration energy, depending on the construction, will reappear at the opposite surface, where it is re-radiated as sound.

The difference between the sound power incident on one side of the partition and that radiated from the second side (both expressed in decibels) is called the sound transmission loss (TL). The larger the sound transmission loss (in decibels), the smaller the amount of sound energy passing through the partition.

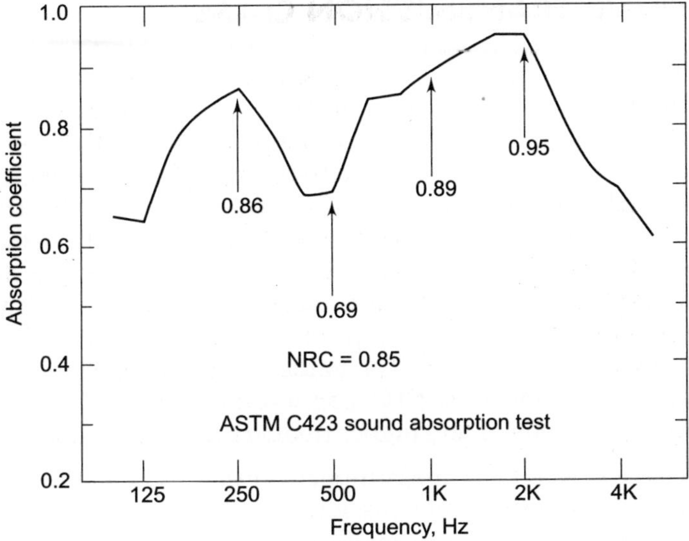

Figure 1.6

The sound transmission loss of a partition generally increases with the frequency of the incident sound and also varies with the direction of the sound waves. When sound transmission loss is measured in a laboratory, the wall or floor under test separates two highly reverberant rooms and ideally is not in solid contact with either of them. Random sound field is generated in one of the rooms and sound energy passes through the partition into the second or receiving room (is shown in Figure 1.7). The difference in the average sound pressure level between the two rooms is called the noise reduction. The level in the receiving room is partly determined by the area of the partition and the total absorption of the receiving room. After correction for these two factors, the noise reduction becomes the transmission loss, which is a property of the partition independent of its size and of the properties of the test rooms. Figure 1.8 gives a plot of transmission loss as a function of frequency.

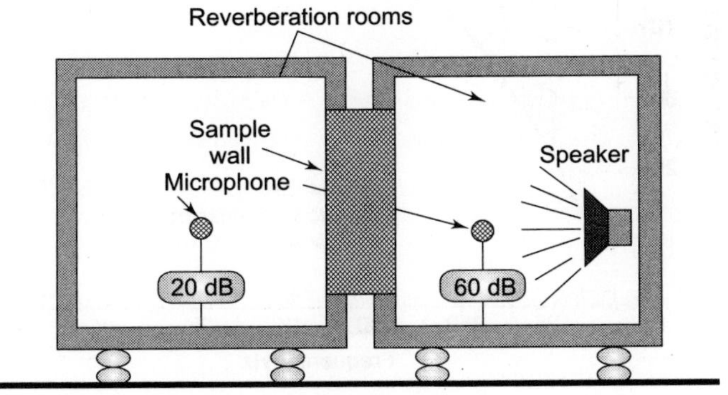

Figure 1.7

1.6   SOUND TRANSMISSION CLASS

As with other acoustical measurements, it is convenient to have a single number rating to describe TL measurements. In North America this rating is the sound transmission class (STC). The larger the STC value, the better the partition, i.e., the less sound energy passes through it. The STC contour is shown fitted to the TL curve in Figure 1.8. The contour extends from 125 to 4000 Hz and the rating is based only on the 16 TL values at these frequencies. Briefly, the reference contour is adjusted until the sum of the differences between it and the TL contour is less than or equal to 32 dB (shown as a shaded area in the drawing). The second constraint is that no deficiency at any frequency can exceed 8 dB. The value of the reference contour at 500 Hz determines the value of the STC. Laboratories often extend the range of the measurements to lower and higher frequencies to get more information.

Sound transmission loss measurements can also be made in buildings and provide a field sound transmission class (FSTC). The letter F simply shows that the measurement was made in a building and not in a laboratory. The field test can be carried out in such a way that all possible transmission paths are included in the measurement. No attempt is made to eliminate alternative transmission paths; obviously the occupant is subjected to the effects of all possible sound transmission paths. The rating so obtained relates directly to the level of privacy enjoyed by the occupants.

To avoid the complexity and time involved in the measurement of sound transmission losses in at least 16 frequency bands, a shorter test procedure has been developed to give a single number rating from simple measurements. It gives an A-weighted level difference between two rooms, corrected for room properties, which is usually within one or two points of

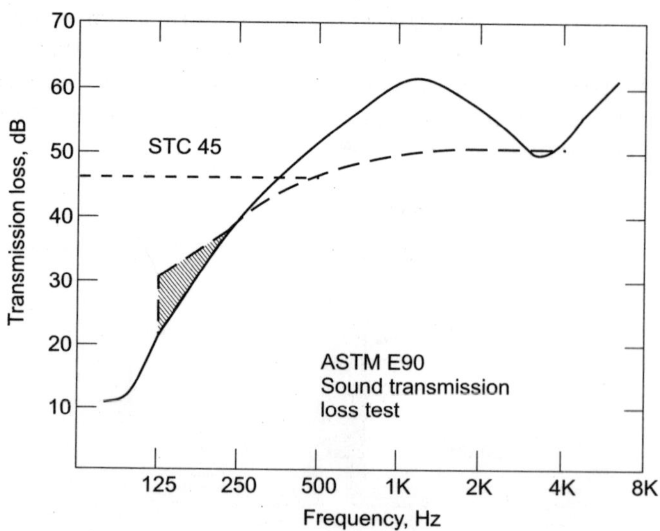

the FSTC values. This can be very useful for quality control in buildings, to ensure that no serious errors have occurred and that the sound isolation is close to the design value.

Impact Noise

Noise from footsteps and other impacts is a common source of annoyance in buildings. A test method in existence for some years gives single number ratings for the transmission of impact sound through floors. This test, originally developed in Europe, uses a standardised tapping machine with five hammers that strike the floor at a total rate often times per second. Sound pressure levels are measured in the room below the floor (is shown in Figure 1.9) in one-third-octave bands and the resulting curve is fitted to a reference contour to obtain a single number rating - the impact insulation class (IIC). An example of this is shown in Figure 1.10, where the shaded portions determine the value of IIC. In contrast to sound transmission loss plots, the higher the measured impact sound pressure levels, the worse the floor. The IIC rating, however, increases as the protection provided by the floor against this specific type of impact increases. The impact insulation class can be determined in the laboratory or in the field. At the moment there is no short impact test equivalent to the short airborne sound test, although one exists in France.

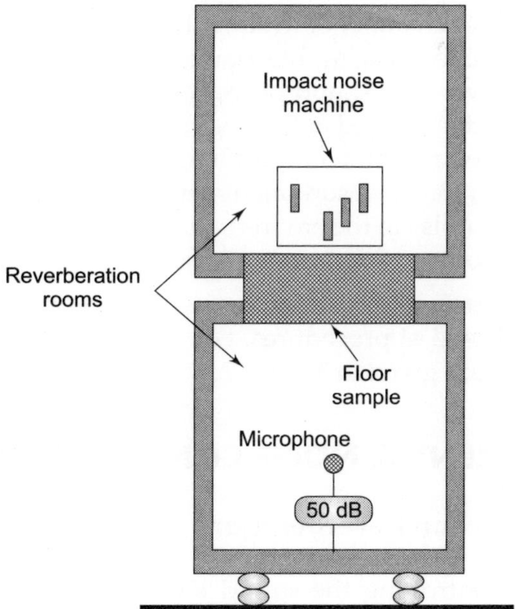

Figure 1.9

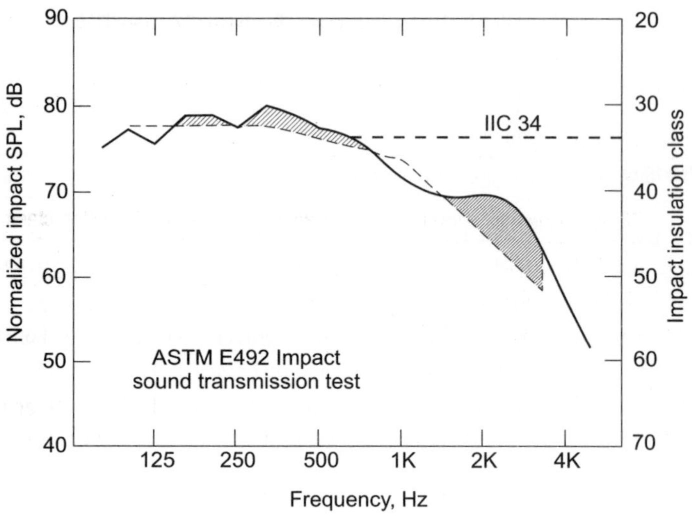

Figure 1.10

Machinery Noise

A variety of laboratory and field test procedures can be used to measure the sound power of machines. The machine can be in a very reverberant test room, a semi anechoic chamber or in situ. The results are usually presented as a graph and a table showing the power in each frequency band and the A-weighted power. The sound power of the machines to be used in a building should be determined at the beginning of an acoustical design. If a quiet, well-balanced machine is specified, there will be less need for the addition of silencers, sound-absorbing materials and high transmission loss partitions to control noise in the environment. It is always most effective to reduce noise at its source whenever possible.

The test procedures described above, with the exception of the sound power tests, are all ASTM procedures. For reference, the test title and the ASTM designation are given in Table 1.2.

1.7   FUNDAMENTAL NOISE CONTROL PROCEDURES

Sound Absorption and Reverberation

One method for controlling the sound level within a room is through dissipation of the sound energy in absorptive materials. Sound is absorbed when a portion of the sound energy striking a surface is not reflected, but passes into the material and is converted into heat energy. Generally, higher

Table 1.2 ASTM Tests used in Building Acoustics

E90	Standard method for laboratory measurement of airborne sound transmission loss of building partitions.
E336	Standard test method for measurement of airborne sound insulation in buildings.
E413	Standard classification for determination of sound transmission class.
E597	Standard practice for determining a single number rating of airborne sound isolation for use in multi-unit building specifications.
E492	Standard method of laboratory measurement of impact sound transmission through floor ceiling assemblies using the tapping machine.
E1007	Standard test method for field measurement of impact sound transmission through floor ceiling assemblies and associated support structures.
E989	Standard classification for determination of impact insulation class.

frequencies are more easily absorbed than low frequencies. Materials that are good absorbers permit sound to pass through them relatively easily; this is why sound absorbers are generally not good sound barriers. They reduce the level of noise inside an enclosure, because while the sound waves are being reflected from the surfaces in the room, they interact with the sound-absorbing materials and lose some energy each time.

However, it requires large thicknesses or many passes for the sound energy to be significantly reduced. It is important to understand the difference between sound-absorption and sound transmission loss. Materials that prevent the passage of sound are usually solid, fairly heavy and non-porous. A good sound absorber is 15 mm of glass fiber; a good sound barrier is 150 mm of poured concrete.

Sound-absorptive materials are used to reduce the level of steady sound in a room, from a machine for example, and to reduce the reverberance. The reverberance or liveness of a room is usually characterized by its reverberation time, the time it would take the energy of an abruptly stopped sound source to decay through a range of 60 dB. Just as sound-absorption varies with frequency, so too does reverberation time. No single figure rating is used to summarize reverberation time information, but the value at 500 Hz is often used for this purpose.

Resilient Mountings

Most machines generate vibration, which can be transferred to the structure of a building through the machine supports. The vibration level can be reduced by balancing the machines to reduce the forces at the source. Transmission into the building structure can be reduced by using resilient, anti-vibration mountings between the machine and the supporting structure. Common materials used as vibration isolators are rubber, cork, various types of steel springs, and glass fiber pads. The transmission of vibrational energy depends on the resonance frequency of the machine on the resilient supports and on the energy damping in the system. The design or selection of resilient mounts is a complicated subject, best left to a competent acoustical consultant, but it is important to know in principle where such devices should be used.

Combination of Techniques

Three important measures are used in almost all noise control work. Figure 1.11 shows a source placed on a resilient support inside an enclosure with high sound transmission losses and lined with sound-absorbing material. If the resilient supports were omitted, the vibrational energy from the machine would pass into the building structure with very little attenuation, to radiate as sound elsewhere. If the solid enclosure were not there, the sound would pass through the sound-absorbing material with very little attenuation. If the sound-absorbing material were omitted, the reverberant sound inside the enclosure would build up and the energy loss through the enclosure would be much less than expected. All three elements should be present.

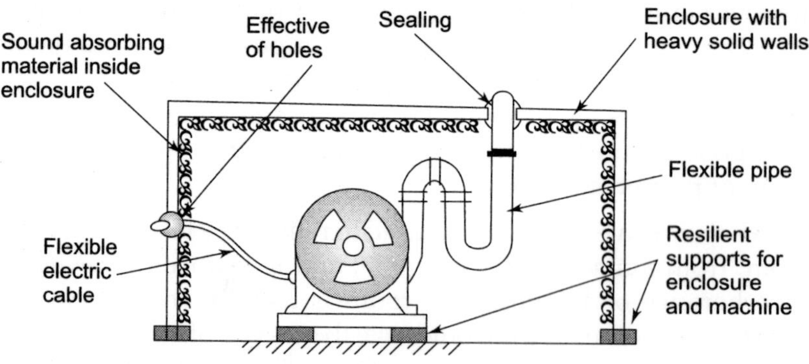

Figure 1.11

The correct application of these three measures is the keystone of good acoustical design and noise control. The enclosure can be a box housing a printer, a large room containing several noisy machines or a living room in an apartment, but the principles remain the same. Good sound control incorporates resilient supports or connections, sound-absorbing material and leak-free sound barriers.

The extent to which these measures are necessary will depend on a number of factors. If quiet machines are selected for use in a building and are placed as far as possible from areas where quiet is required, fewer noise control procedures will be necessary. It is not always possible, of course, to control the sources of noise in a building, since sometimes the occupants are the sources.

2

SOUND IN ROOMS

2.1 INTRODUCTION

Rooms are usually intended for specific people-related tasks. The acoustical conditions in rooms must be such that the intended activities are optimally supported and are not in any way hindered by the room. Two basic aspects govern the acoustical conditions in rooms: (1) the amount of background noise, and (2) the acoustical properties of the room itself, as determined by the geometry and materials of the room.

Background noise levels must be limited to some suitable maximum value depending on the intended use of the room, which may vary from sleeping, to listening to a person speaking, to industrial activities. Room acoustics are conventionally evaluated in terms of obtaining an optimum reverberation time. Each type of use has different acoustical requirements in terms of both background noise levels and room acoustics conditions.

There are three steps to the acoustical design of rooms: (1) selecting optimum criteria to be achieved dependent on the intended use, (2) designing to achieve these criteria, and (3) verification that the design goals are achieved in the finished building. All three steps are essential to achieving a successful room. While the building designer must determine the optimum criteria for each room, the support of mechanical and acoustical consultants will be required to ensure that an acoustically successful design is achieved. Although critical design details and the particular problems of rooms such as auditoria require the knowledge and experience of specialists, the basic principles that apply to many types of rooms are quite straightforward.

2.2 BASIC ACOUSTICAL PRINCIPLES

Sound Power and Sound Pressure

One can readily measure sound pressure with a microphone and associated electronics. The measured sound pressure is related to how loud the sound is perceived to be, and will vary with position relative to the source and with the acoustical conditions in the room. Thus a particular noise source would be measured as producing different sound pressures in different rooms, as it would sound louder or softer in different rooms. The sound pressure is not simply a property of the source but is determined by a combination of the source properties, the room properties, and the source-receiver distance. The sound power of a source, on the other hand, is essentially constant and hence is only a property of the source.

An analogy can be made between a noise source and a light bulb. A light bulb is rated to dissipate a particular number of watts of power. The bulb will always dissipate the same amount of power independent of its surroundings. However, the same bulb may appear to brighten a room covered with shiny reflecting walls more than another with dull black walls. Similarly the bulb might be adequate in a small room, but in a large room might appear very feeble. This is an exact parallel to the acoustical situation.

It is much more useful to know the amount of sound power produced by a source than to know the sound pressure measured under some particular condition; one should strive to obtain sound power ratings of all potential noise sources at the design stage. One can then calculate the resulting sound pressure levels in any room if the source sound power levels and some properties of the room are known. Equation (2.1) in Appendix 2.1 can be used for this purpose.

Steady State Sound Levels in Rooms

Noise levels in rooms vary with distance from the source and with the properties of the room. In an outdoor situation (in the absence of any reflecting surfaces), sound levels decrease as one moves away from the source. Sound pressures decrease inversely proportional to the distance from the source, that is, sound levels decrease 6 dB for each doubling of distance from the source. This is true for a source that radiates sound approximately equally in all directions and where there is only one path from the source to the receiver. In a room, there are a very large number of possible paths from the source to the receiver, involving various reflections off the room boundaries; the combination of all these paths determines how sound behaves in the room. As a result, sound levels in a room do not continue to decrease with increasing distance from the source for all distances.

Figure 2.1 gives some examples of how sound levels vary with distance from a source in a typical 2000 m^3 room. Although sound levels initially decrease with increasing distance from the source, as they would in the absence of reflections, after some distance from the source, sound levels become relatively constant as distance is further increased. Closer to the source, sound levels depend on the properties of the source and the distance to the source, but further away, sound levels are dominated by the reverberant sound energy, which depends on the properties of the room. The harder and more reflective are the room boundaries, the more reverberant sound energy there will be and the higher will be the reverberant levels in the room. Conversely, for a given room boundary material, reverberant field sound levels will decrease as the surface area is increased because more reverberant energy will be absorbed.

Figures 2.1(a) and 1(b) are simple examples of noise control strategies in a 2000 m^3 room (30 × 15 × 4.5 m). Figure 2.1(a) illustrates the effect of adding sound-absorbing material to the room. The upper solid curve corresponds to a room where all the boundaries are quite hard and reflective (absorption coefficient equal to 0.1 and total absorption equal to 130 metric sabins). For this case, the reverberant field sound level is relatively high. If the total absorption is increased to 450 metric sabins (by making the ceiling highly absorptive), the dashed curve results and the expected reverberant field level is reduced by over 5 dB. If total absorption in the room is increased to 800 metric sabins, as would be the case if the ceiling and all walls were covered with highly absorptive material, the dash-dot curve on Figure 2.1(a) results, with a further small reduction in the reverberant field levels. This demonstrates that going from all hard reflective surfaces to all walls and the ceiling being highly absorptive, leads to a maximum of 8 dB reduction in the reverberant sound levels. In practical situations, it is difficult to achieve such a large relative change. As the figure also demonstrates, much smaller reductions are produced closer to the source, and treating the room is only effective at positions remote from the source in the reverberant field. [Equation (2.2) in Appendix 2.1 allows one to calculate the reduction in reverberant field sound levels due to the addition of absorption.]

Because a screen can reduce the direct sound close to the source, and added absorption can reduce the reverberant sound levels farther from the source, the combination is potentially useful to generally reduce the sound levels throughout a room. Figure 1b also illustrates such a combination for a highly absorptive ceiling (as in Figure 2.1(a)), and a screen that reduces the direct sound by 10 dB. The result is a reduction of 5 dB or more throughout the room. These examples illustrate the reductions that are likely in realistic situations; very large reductions in noise levels cannot be achieved by treatment of the room. It is essential to control noise at the source by selecting devices with adequately low noise output. Quieting the source later is usually very difficult and special noise reducing enclosures are often limited by practical considerations such as the need for access to the noise-producing device.

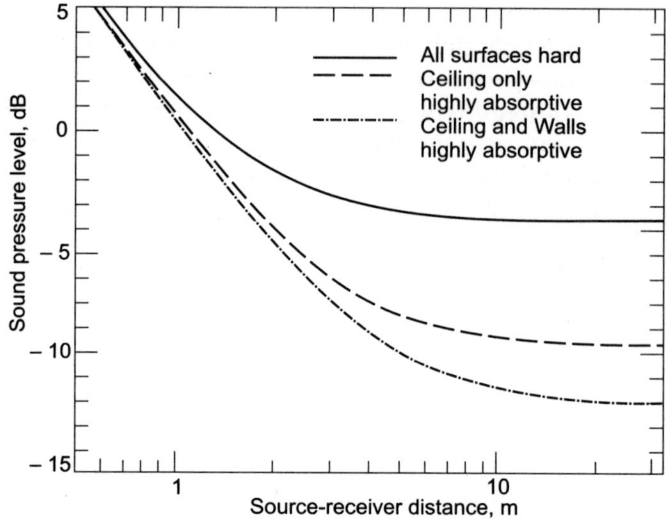

Figure 2.1(a)

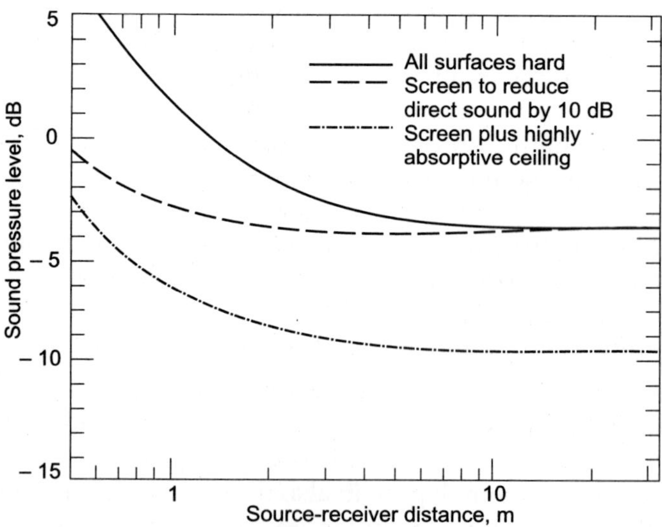

Figure 2.1(b)

An alternate noise control strategy is to use a screen to block the direct sound. Figure 2.1(b) illustrates the estimated effect of a screen that reduces the direct sound by 10 dB. Close to the source the sound levels are reduced by several decibels, but there is no change to the reverberant sound levels farther from the source.

Sound-absorbing Materials

Sound-absorbing materials are described by their sound absorption coefficient measured over a range of frequencies. Porous sound-absorbing materials can be made of glass fibre, rock wool, open cell urethane foam, cloth or other materials that are porous to air flow. They are characterized by high absorption coefficients at high frequencies, decreasing at lower frequencies depending on the type and thickness of the material. Figure 2.2 illustrates typical absorption coefficient values versus frequency for 2.5 and 5.0 cm thick porous samples. Such absorbing materials function by resisting the air flow associated with the acoustical vibrations of the air, and are most effective at higher frequencies, where they are thicker relative to the wavelength of the sound. Since absorption decreases at lower frequencies, porous absorbing materials a few centimetres thick will never be highly absorptive at lower frequencies. As Figure 2.2 demonstrates, increased low frequency absorption can be obtained by adding an air space between the material and the rigid backing so that the sound absorber behaves like a thicker material.

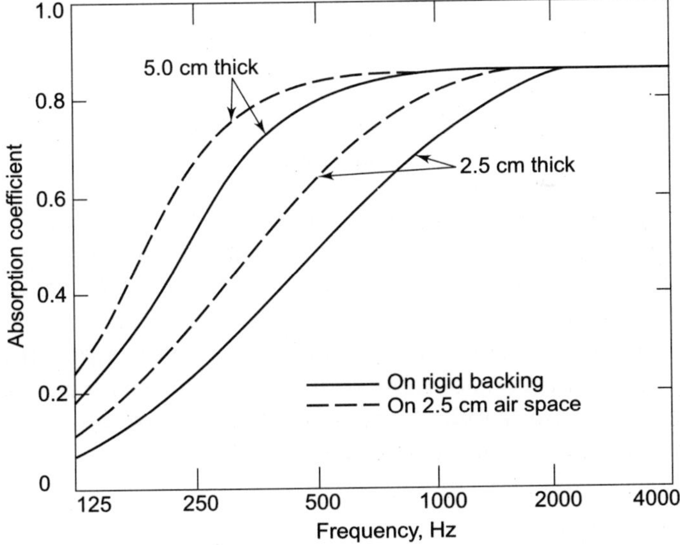

Figure 2.2

Where low frequency sound absorption is required, one of several types of tuned absorbing systems must be used. Figure 3 illustrates the absorption possible from a membrane absorber consisting of a thin panel with an enclosed air space. This constitutes a simple resonant system, where the mass per unit area of the panel and the stiffness of the enclosed air space determine the frequency at which the absorption is maximum. As the depth of this air space is increased its effective stiffness is decreased. [See also Equation (2.3), Appendix 2.1.]

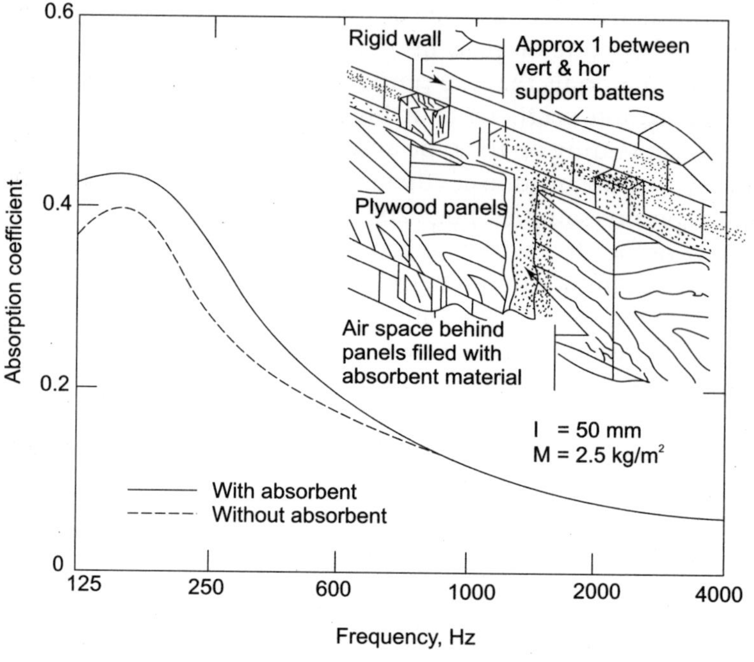

Figure 2.3

Similarly, cavities with a small opening to the room can be tuned to a particular frequency at which they will absorb maximally. Figure 2.4 illustrates the absorption characteristics of such a resonant cavity absorber. The same principle is used in commercially available concrete blocks with slits. Such blocks, when combined with a layer of porous absorbing material, can provide considerable sound absorption over a broad range of frequencies. Equation (2.4) in Appendix 2.1 can be used for predicting the resonant frequency of a tuned cavity absorber.

Reverberation time

In many types of rooms it is important to optimize the reverberation time. This is approximately the time required for a loud sound to fade away to inaudibility after the source has been turned off. More precisely, it is the time required for the sound level to decrease by 60 dB after the source has been abruptly stopped. Reverberation time is directly related to the room volume and inversely related to total absorption in a room. [See also Equation (2.5) in Appendix 2.1]. Thus larger rooms tend to have longer reverberation times and are said to be more acoustically 'live' or reverberant. Smaller rooms and rooms with more sound-absorbing material will be less reverberant and may be described as acoustically 'dead'. There is also a connection between increased reverberation time and increased reverberant sound level, as both quantities are related to the total amount of sound-absorbing material in a room. Thus attempts to control reverberant field noise levels by the addition of absorbing material will simultaneously reduce reverberation times. In rooms intended for speech (and especially in rooms used for music), one must design for an optimum reverberation time. Too much reverberation causes one word to blur into the next and hence reduces the intelligibility of speech. Too little reverberation leads to a reduction in reverberant field speech levels. (The design of rooms for speech is considered in more detail in a later section.)

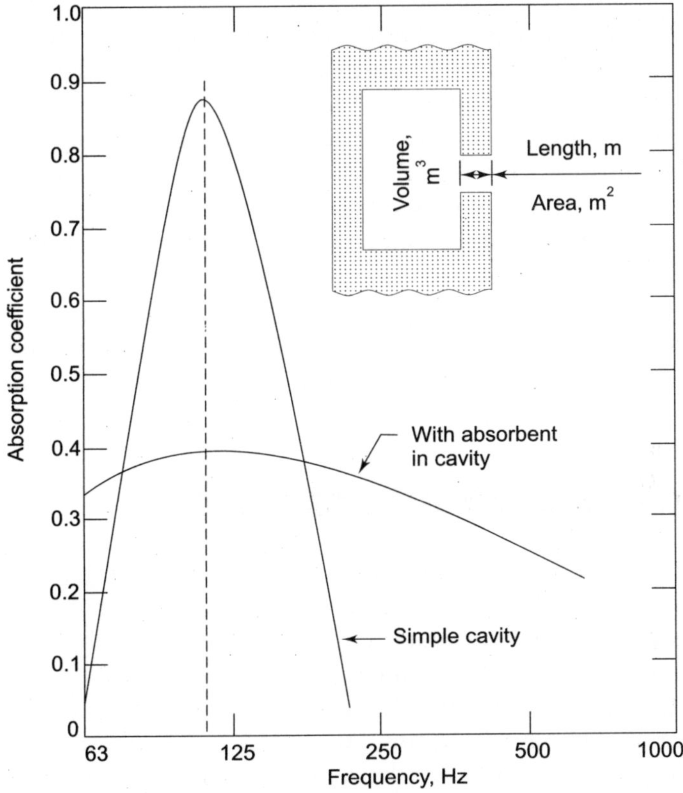

Figure 2.4

2.3 NOISE CRITERIA FOR ROOMS

Although a number of procedures have been developed for rating the acceptability of noise in buildings, two are adequate for most situations. The simplest is to measure directly overall A-weighted sound levels. Where a more detailed procedure is required, the noise criteria (NC) rating procedure, requiring measurements in octave bands, is normally used (see *in Chapter 1*).

An acceptable level of background noise in a room is one that does not interfere with the activities intended for the room. In large industrial spaces, noise limits may be based on the relatively high levels tolerated by occupational noise regulations. At the other extreme, in a recording studio, almost any audible noise is undesirable. For many rooms, such as classrooms and meeting rooms, intended for speech communication, quite low levels of background noise will interfere with the precision and quality of this communication.

Table 2.1 summarizes optimum maximum acceptable noise levels for various activities. The highest limits in this table correspond to occupational noise exposure limits of 85 or 90 dBA for an eight-hour working day. These are intended to protect workers from permanent hearing loss. As background levels decrease from 75 to 35 dBA, the quality of speech communication, and the acceptable distance between talker and listener gradually increase. At 75 dBA, shouted warnings, often necessary for worker safety, would be audible over a distance of a few metres. With a background of 35 dBA or lower, very high quality speech communication should be possible in rooms similar in size to school classrooms. As the room volume increases, lower background noise levels are required (this is discussed in more detail in the following section).

2.4   DESIGN OF ROOMS FOR SPEECH

A large range of rooms are intended almost entirely for groups of people to listen to speech. These vary in size from small meeting rooms and classrooms to large auditoria and conference centres. Where rooms have more than one use, the criteria for excellent speech conditions will be more restrictive than those for many other activities. In larger auditoria, theatres, and other more complex situations, an acoustical consultant is essential to a successful design. For many smaller rooms, optimum conditions for speech can be achieved by the combination of an acceptably low background noise level, and an optimum reverberation time. Early reflections are particularly important to good speech intelligibility; in designing the geometry and materials of the room, the attaining of strong early reflections must be considered.

Table 2.1 Some Optimum Maximum Noise Levels Acceptable for Various Activities

Maximum Levels (dBA)	Activity	Examples
85 or 90	Occupational noise limits (per 8 hour day)	Factories and industrial settings
75	Shouted warnings audible within 3 to 4 metres	Industrial locations
65	Acceptable face – face communication	Workshops, garages, plant control rooms
55	Non-critical speech for nearby listeners	Public areas of buildings, lobbies, restaurants
48	Maximum acceptable masking sound level	Open plan offices
40	Good speech quality for groups of nearby listeners, tasks involving high concentration	Private offices, libraries
35	High quality speech for groups of listeners (decreasing with room volume)	Classrooms, meeting rooms
30	Sleep, high quality speech for special sensitive groups	Bedrooms, classrooms for very young or hearing impaired
25	High quality unamplified speech, speech in larger rooms	Concert halls and theatres

Figure 2.5 provides a simple design contour describing optimum maximum background noise levels in rooms for excellent speech intelligibility as a function of room volume. The background noise level in rooms intended for speech should never exceed 35 dBA, and larger rooms require lower background levels. The design contour was derived from the individual lines for various speech source levels ranging from a 'normal' to a 'raised' voice level. Thus in Figure 2.5 the lowest dotted line represents a 'normal' voice level, while the top dotted line represents a 'raised' voice level. In smaller rooms one expects to be able to talk in a 'normal' voice, while in much larger rooms one reasonably expects to have to raise one's voice. Thus the design contour (solid line) represents the combination of these effects for a range of room volumes. In particular cases, larger rooms may be designed so that high quality speech is maintained at lower speech source levels, such as in a theatre, where the full dramatic intent must be conveyed.

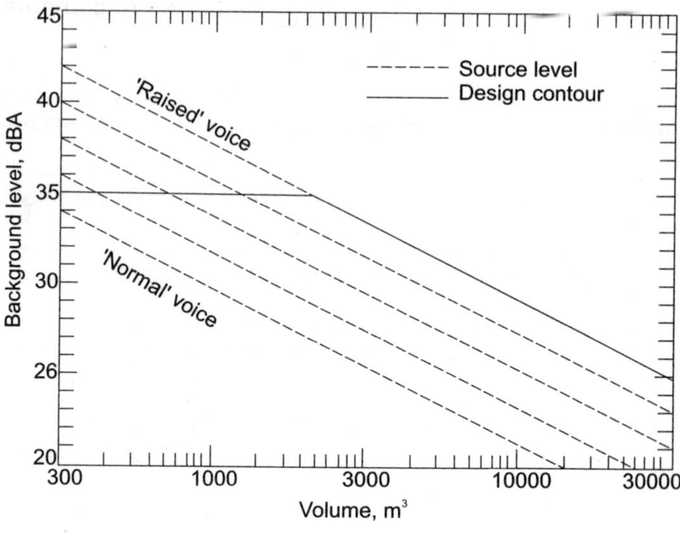

Figure 2.5

Figure 2.6 plots optimum reverberation time for speech versus room volume. As the room volume increases, the optimum reverberation time also increases. Since this contour indicates a unique optimum reverberation time for each room volume, it is possible to calculate the related optimum total absorption necessary for each room volume. Figure 2.7 shows the resulting range of acceptable total room absorption (in metric sabins) versus room volume (in cubic metros) that would lead to the desired optimum reverberation time within ± 0.1 second. Designing for a total room absorption as close as possible to the middle of this range should lead to nearly optimum conditions for speech.

Particular groups of listeners require more restricted acoustical conditions to obtain optimum speech intelligibility. These groups include: children of elementary school age, retirement age adults, and all hearing impaired persons. For such groups one should consider a further 5 dBA reduction in background levels and ensure that optimum reverberation times are not exceeded.

The geometry and surface materials of a room for speech must be selected to ensure that strong early reflections are provided; these should arrive within approximately 0.05 seconds after the arrival of the direct sound. In other words, the lengths of these reflected sound paths should be no more than 15 metres longer than the direct path from source to receiver. These early reflections are fused with direct sound by the hearing system and lead to increased speech intelligibility. Sound reflections arriving later than this are detrimental to speech intelligibility. Newer acoustical measures have been devised to measure the relative benefits of these early reflections, and some new measures even successfully combine measures of the background noise and the room acoustics into a single quantity. These are not yet in widespread use. In many medium to smaller rooms one can reasonably design for an optimum reverberation time and qualitatively design for increased early reflection energy.

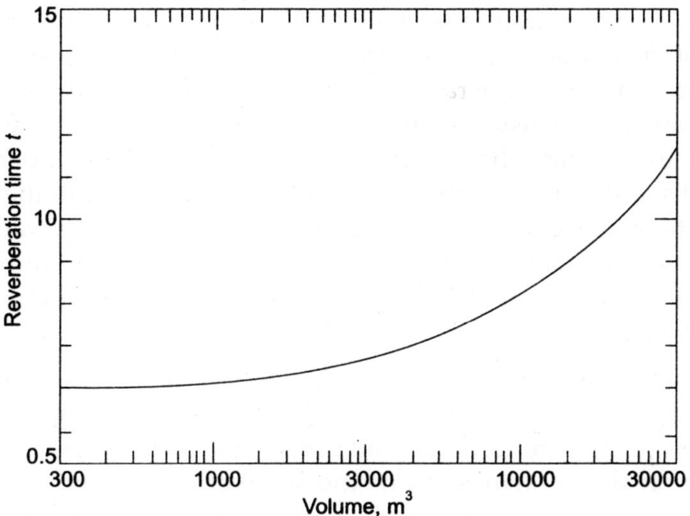

Figure 2.6

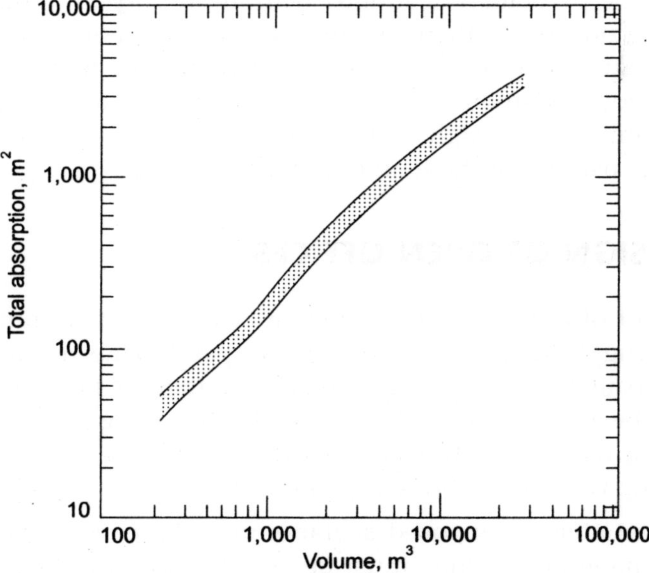

Figure 2.7

Specific recommendations to qualitatively design for increased early reflection energy would vary somewhat with room size, and the mode of use of the room. In smaller rooms, such as classrooms, a person may wish to speak from any point within the room. While strong early reflections are desired, Figures 2.6 and 2.7 indicate that such small rooms should have a quite short reverberation time of about 0.7 seconds and hence require the addition of some sound-absorbing material. The problem is where to add the material so that the reverberation is controlled, but strong early reflections are maintained. The usual strategy would be to ensure that the central portion of the ceiling is hard and reflecting, as well as the upper portions of the walls, if this are possible. Absorption is best added to the floor, the lower portions of the walls, and the cornice area between the walls and the ceiling. Absorbing material added in the cornice area by mounting sloping panels from the ceiling to the wall provides absorption over a broad range of frequencies.

In larger rooms there is usually a source end, with a stage or platform where the speaker is normally located. It is necessary to design this area to provide strong early reflections to all parts of the audience, with particular emphasis on those farthest from the speaker. It may be necessary to make the rear wall either diffusing or absorptive to prevent long delayed reflections that could be heard as echoes. Where a medium or larger sized room is intended almost entirely for speech, the services of a good acoustical consultant are essential to ensure that a high quality room results. One should be particularly cautious with multi-purpose rooms where a considerable portion of their use involves speech. For example, school gymnasia frequently receive no acoustical forethought. In practice they are used as a school classroom and often serve as a school auditorium, as well as a community hall at night. Ignoring the acoustical design means they may be quite useless for these other purposes.

2.5   DESIGN OF OPEN OFFICES

Open plan of offices are a very different acoustical environment, where one must struggle to prevent the hearing of speech. Thus the design goal is exactly opposite to that in rooms for speech. Attempts to achieve this goal will at best be a compromise. One must do everything possible to absorb and to block the sounds from one work area so that they do not reach neighbouring work areas, but without the aid of walls.

As a first step all exposed surfaces must be covered with sound-absorbing material. Ceilings must be very highly absorbing and floors must be carpeted. Direct sound paths must be blocked where possible with screens that both act as a barrier to sound and have sound-absorbing surfaces. Screens should have an STC of at least 20 (see in Chapters 3,4);

this requires an impermeable central layer of at least 2.4 kg/m^2. Screens should be at least 1.5 × 1.8 m in size and should extend to the floor and other adjacent surfaces. Any large exposed wall areas should also be treated with absorbing material.

Even when such treatment is carried out very thoroughly, acoustical privacy between adjacent work areas will not be complete. Electronic masking noise systems are often added to provide low levels of neutral broad band noise to mask speech and other sounds from neighbouring work areas. Such masking noise will only be successful if it has a neutral spectrum, is maintained at a constant sound level at all times and evenly throughout the office within ± 2 dB, and if the level of masking noise is not too loud. Masking levels greater than 48 dBA lead to occupants raising their voices and to increasing levels of annoyance.

When the acoustical design is carried out without compromise, the space can function for activities not requiring considerable concentration. Particular problems of inadequate acoustical privacy may be encountered when lighting fixtures, windows, or furniture provide unwanted reflected sound. The heavily treated open office is not a semi- reverberant room, and reverberant sound levels as illustrated in Figure 2.1 do not exist. Similarly, it is meaningless to attempt to consider the reverberation time of such a space.

The design procedure followed for open plan offices does not apply to open concept schools, where it is again essential to hear speech. The concept of an open plan school is completely inconsistent with the goals of good design for optimum acoustical conditions for speech.

HVAC Noise

In the majority of rooms the most common source of disturbing noise is mechanical system noise or HVAC noise. While this may be a very common problem, there is usually no excuse for it because detailed design procedures such as those in the ASHRAE guide have existed for many years and are frequently upgraded. HVAC noise problems appear to arise more from ignoring the proper design process than from inadequate technology. Because such design procedures involve detailed consideration of the HVAC system, they will not be repeated here, but the basic principles of this design process will be presented so that building designers can be aware of the general concepts.

One must start with known sound power levels for each system fan, which should be selected for its low sound power levels. Following the calculation procedures, one can determine how much of this sound power enters each room that the fan supplies. In each room the sound power levels are converted to octave band sound pressure levels according to the acoustical properties of the room, and NC values are calculated. These are compared to the established noise criteria for the room, and where the criteria are exceeded, additional steps are taken to further attenuate the HVAC noise. It is important to appreciate that the sound energy can travel by several different paths. The vibrational energy of the fan can propagate through the fan room floor and other parts of the building structure, as well as through the walls of the duct system. The airborne sound energy from the fan can propagate through the duct system in both directions from the fan, as well as into the fan room from the fan casing. All of these paths can be very important and one cannot ignore any one of them. Clearly the inappropriate location of potentially noisy mechanical equipment can lead to severe noise problems; it cannot simply be tucked into any unused corner of the building.

The overall design procedure is to some extent iterative; if, after an initial calculation, noise criteria are exceeded, changes must be made and a further set of calculations must be carried out. The procedure can be described in overview by the following eight steps.

1. Select a suitable location for mechanical equipment rooms away from noise-sensitive areas of the building.

2. Select fans with low sound power output levels. Ideally the octave band sound power levels should be known for both the sound radiated into the duct and that from the fan casing into the fan room.

3. Vibrationally isolate the fan. This includes both isolating the fan and motor from the floor, and installing a vibration break in the duct immediately adjacent to the fan, to prevent the propagation of vibrational energy through the duct walls.

4. Acoustically isolate the fan noise. The equipment room walls, floor, and ceiling must provide a high transmission loss to the airborne noise in the equipment room (see *in Chapters 3, 4* for the requirements of such walls). One must attenuate the sound propagating down the duct by the installation of a dissipative silencer immediately after the duct vibration break. Such a silencer is selected to provide adequate attenuation in each octave band and also to not produce excessive airflow noise.

5. Calculate the attenuation of the sound propagating through the duct system so that the sound power entering each room is known.

6. Estimate flow noise sound power levels entering each room. Sharp bends, control mechanisms, grills, diffusers, and combinations of flow-noise-producing devices placed too close together will lead to increased flow-induced noise. Flow-induced noise is very dependent on the velocity of the air flow and can thus be dramatically reduced by reducing this velocity.

7. Calculate the combined sound power entering the room and the expected octave band sound pressure levels in each room. From these, calculate an NC value and compare it with the established noise criteria for the room.

8. If the desired noise criteria are exceeded, further reductions are needed. If fan noise propagating through the duct is the dominant source of noise in the room, an improved dissipative silencer would help. If flow-induced noise is the dominant problem, lining the last few metres of the duct with sound-absorbing material would help. This would require a slightly larger duct to maintain the original internal dimensions and flow velocities. Reducing the air flow velocity or modifying particular flow-noise-producing devices could also be considered.

2.6   APPENDIX 2.1 ROOM ACOUSTICS EQUATIONS

Sound Levels in Rooms

The relation between sound pressure and sound power levels in a semi-reverberant room can be calculated with the equation below. This equation is based on the assumption of a diffuse sound field in which, at all points in the room, equal amounts of sound energy travel in all directions. Where the room contains large amounts of sound-absorbing material, or where the room is much longer in one dimension than in the other, the equation would become increasingly less accurate.

$$\mathrm{SPL} = \mathrm{SWL} + 10 \log\{(Q)/(4\,\pi r^2) + 4/A\} - 0.2, dB \qquad ...(2.1)$$

where

SPL is the sound pressure level, in decibels, relative to the standard reference pressure of 0.00002 Pa;

SWL is the sound power level, in decibels, relative to the standard reference power of 10^{-12} W;

Q is the directivity factor of the source in a particular direction from the source. It is the ratio of the energy travelling in this direction to the average over all directions. For a source radiating equally in all directions, Q would be equal to 1.0;

r is the source–receiver distance in metres;

A is the total sound absorption, in metric sabins, and is obtained from the product of the mean absorption coefficient and the total surface area. The total absorption in a room can be calculated as the sum of the adsorptions of each surface, where the absorption of each surface is the product of the surface area and its absorption coefficient.

From Equation (2.1), one can calculate the expected reduction in reverberant field sound levels as a result of adding absorption to the room. This can be done more simply by the following equation,

$$SPL = 10 \cdot \log \{Ab/Aa\}, \text{dB} \qquad \qquad \text{...(2.2)}$$

where:

SPL is the change in sound level,

A_b is the total sound absorption before the change,

A_a is the total absorption after the change.

Resonant Sound Absorbers

The resonance frequency of a tuned membrane absorber such as a thin plywood panel over an enclosed air space can be calculated from the following equation:

$$f = 60/\sqrt{ML} \qquad \qquad \text{...(2.3)}$$

where:

M is the mass per unit of surface area, in kilograms per square metre, of the membrane,

L is the depth of the enclosed air space, in metres.

The resonance frequency of a tuned cavity absorber is given by the following:

$$f = 55/\sqrt{s/(l \cdot V)} \qquad \qquad \text{...(2.4)}$$

where:

s is the cross sectional area of the neck opening, in square metres,

l is the length of the neck, in metres,

V is the volume of the cavity, in cubic metres.

Such resonant absorbers would be expected to absorb maximally at their resonance frequency, as illustrated in Figures 2.3 and 2.4.

Table 2.2 Values of 4 m, Four Times the Energy Attenuation due to Air Absorption, at 20°C.

Relative humidity, %	Frequency, Hz		
	2000	4000	8000
30	0.0119	0.0379	0.136
50	0.0096	0.0244	0.086
70	0.0085	0.0213	0.060

Reverberation Time

Reverberation time can be related to the volume and total sound absorption in a room by the Sabine reverberation time equation.

$$RT = 0.161 \, V/(\alpha \cdot S + 4 \, mV) \qquad \qquad ...(2.5)$$

where:

V is the room volume in cubic metres,

α is the mean absorption coefficient,

S is the total surface area of the room, in square metres,

m is the energy attenuation constant per metre due to air absorption. Values of 4 m at 20°C are given in the table below for various combinations of relative humidity and frequency. At lower frequencies, air absorption can be neglected and m given a value of zero.

3

DEVELOPMENTS IN NOISE CONTROL

3.1 INTRODUCTION

Unless requirements are laid out in codes, control of noise in buildings is often an afterthought. The measures taken to control noise, however, are invariably linked to other building subsystems. Mechanical and plumbing subsystems generate noise; the design of walls, ceilings and floors affects sound transmission.

This chapter addresses four topics:

- sound transmission through concrete blocks
- plumbing noise
- flanking noise
- noise leaks.

It assumes a certain background in acoustics and will explain briefly only those terms and ideas relevant to the topics under discussion. Readers not familiar with some of the basics of noise control are referred in *Chapter 1*.

3.2 SOUND TRANSMISSION THROUGH BLOCK WALLS

Background-Transmission Loss (TL) and Sound Transmission Class (STC)

Transmission loss (TL) is the loss in sound power that results when sound travels through a partition. The more power that is lost, the greater the

TL. Figure 3.1 shows sound transmission loss values for some common materials. For single layers of common materials, TL values range from about 10 to about 80 dB.

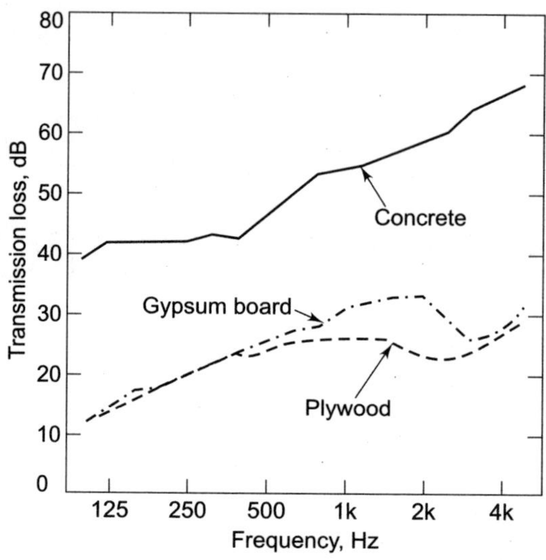

Figure 3.1 Sound transmission loss through building materials

TL depends on frequency; it generally increases as frequency increases. Low frequencies pass through walls much more easily than high frequencies. That is why bass guitar and drum sounds from adjacent apartments are usually most prominent; they are mostly low frequency. HVAC systems may also contain a great deal of low frequency sound.

The smallest difference that people can detect easily is about 3 dB. There is little point therefore in worrying about TL changes of one or two dB.

TL plots for building materials present too much information to be easily assimilated. It is more common to use the sound transmission class (STC). STC is a single number rating that summarizes airborne sound transmission loss data. Figure 3.2 shows the STC contour fitted to the TL curve for concrete block. Once the fit is carried out according to the roles laid down in the ASTM standards the STC value is read from the reference contour at 500 Hz. The higher the rating, the more sound is blocked.

The 1990 National Building Code of India (NBCI) requires an STC rating of 50 for party walls and floors. This is an increase of 5 dB over previous Code requirements. Acousticians, however, usually recommend a design STC of more than 50, say 55 or 60. There are several reasons for this:

- Constructions often perform less well in buildings than they do in laboratory tests. A higher component design rating gives a better chance of meeting Code and overall system requirements; it provides a margin of safety.

- The higher the STC, the less chance there is that building occupants will complain. The higher quality does not necessarily increase costs.

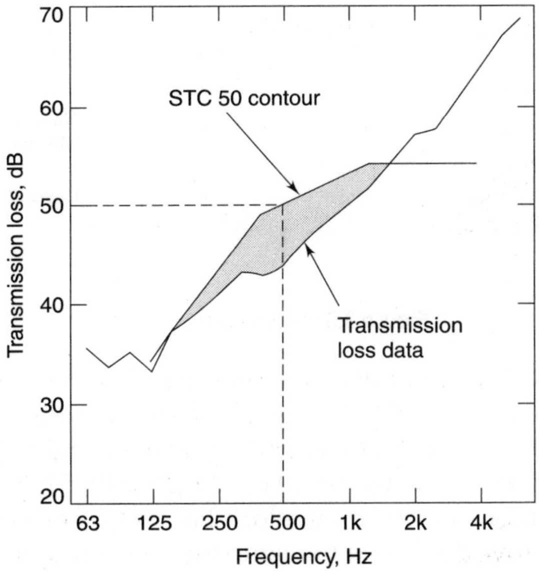

Figure 3.2 STC contour fitted to TL curve for concrete block

- As shown in Figure 3.2, the STC fitting procedure only extends to 125 Hz. Thus, walls may be quite weak below 125 Hz, yet this excessive transmission at low frequencies may not influence the STC rating.

Designers will not be aware of these weaknesses if they only look at STC ratings. Low frequency noise can be a great problem.

Specifying higher STC ratings provides some protection against poor low frequency sound insulation. In fitting the STC contour to the TL curve, the 8 dB rule states that no TL value can be more than 8 dB below the STC curve. Thus the STC is unlikely to be high when the low-frequency transmission loss is very poor. An example of the protection this rule provides is shown later in Figure 7.

3.3   INCREASING THE STC OF BLOCK WALLS

One of the most effective ways of increasing sound transmission loss is to use double layer construction, that is, two layers of material separated by an air space (Figure 3.3). Increasing the weight of the layers in a double wall, increasing the depth of air space, or adding sound-absorbing material, all increase the transmission loss and therefore the STC rating for a double wall. There should be no solid connections between the two layers. Resilient connections, such as those provided by resilient metal channels or non-load-bearing steel studs, are acceptable in most cases.

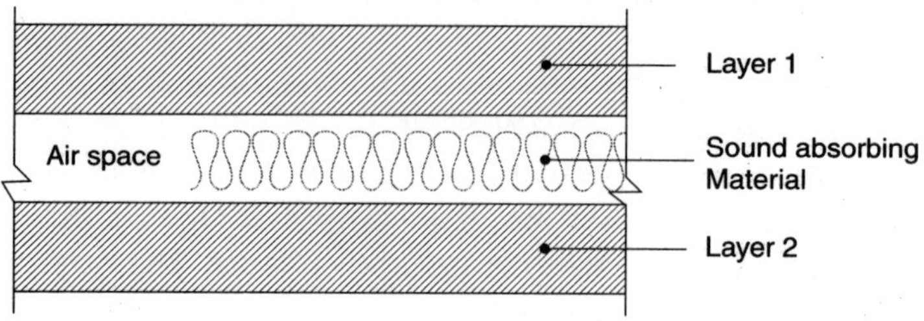

Figure 3.3 Idealized wall section

Concrete block is a popular building material that offers fairly good sound insulation because of its weight. Normal weight block (19 cm) provides about STC 50, or less, depending on the weight of the block. This is not quite good enough to be sure of meeting the 1995 NBCI requirements; in any case, in home or office the block has to be finished, usually with drywall. To improve the sound transmission loss through block walls, one can support the drywall away from the block to form a double or triple layer wall. It is important to know just what effects one can expect with different methods of attaching drywall. What STC ratings can be achieved? What happens at low frequencies?

At NPL we looked at different ways of attaching drywall to block walls to answer these questions (as shown in Figure 3.4). With the exception of the wood strapping, all of the supports were resilient. Walls were tested with and without glass fibre batts in the cavities and with one or both sides finished.

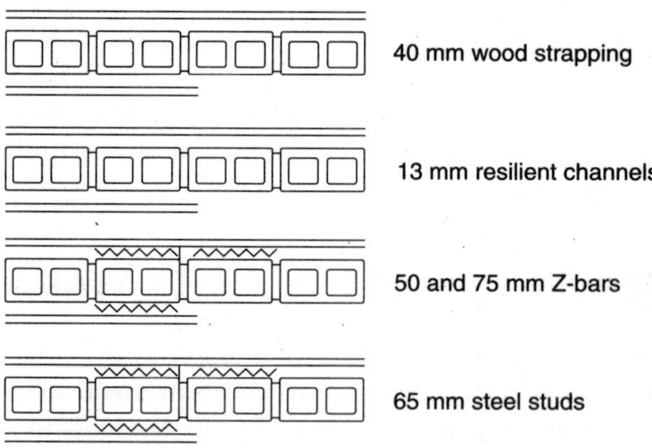

Figure 3.4 Wall assemblies tested for sound transmission loss

Figure 3.5 presents the results for bare block, for the block with 16 mm drywall supported on 13 mm resilient channels, and for 16 mm drywall supported on 75 mm z-bars. The curves for the walls with added drywall fall below the curve for the bare wall at the left, or low-frequency, end of the graph. This is caused by a resonance between the drywall and the air in the cavity. The air acts as a spring and the drywall bounces on it, much like a ball bouncing on a piece of elastic. The larger the air cavity or the heavier the drywall, the lower the frequency where the resonance occurs. This resonance is called the mass air-mass resonance. The first mass is the drywall, the second is the block, which is so heavy that it has little influence on the position of the resonance.

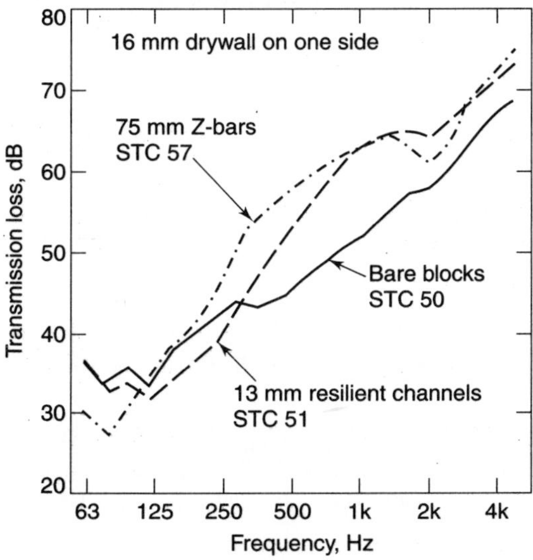

Figure 3.5 Sound transmission loss through different wall assemblies

Figure 3.6 compares the results for a 50-mm deep cavity with and without glass fibre in it. The sound absorbing material makes the cavity respond as if it were about 40% larger and the resonance moves to a lower frequency. In general, when sound absorbing material is added to a cavity, the transmission loss improves and, if the resonance frequency is low enough, the STC usually increases.

Table 3.1 summarizes the results where drywall is attached on one or both sides of the wall. As the air space increases, the STC goes up. Treating both sides also usually increases the STC. The highest value obtained was STC 72. There are, however, one or two peculiar results. Adding drywall, resilient metal channels and glass fibre on both sides of the wall caused the STC to drop one point.

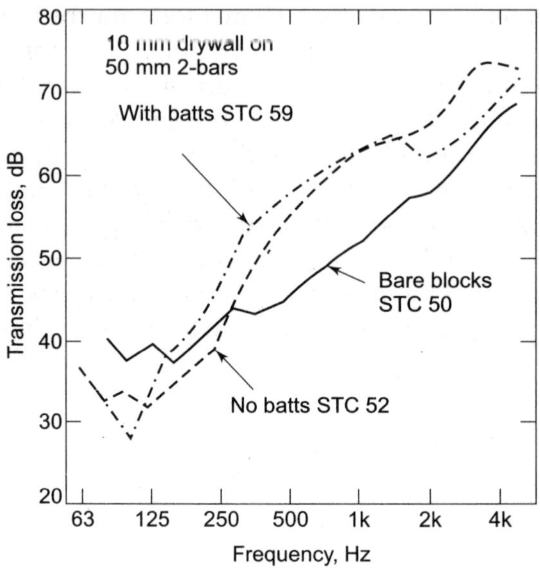

Figure 3.6 Sound transmission loss through walls with and without insulation in the cavity

Table 3.1 STC ratings for 190 mm normal Weight Block Walls, with Different Methods of Mounting 16 mm Drywall with and without Glass Fibre Batts Filling the Cavities

	No glass fibre		With glass fibre	
Drywall attachment	One side	Both sides	One side	Both sides
Bare blocks	50			
Applied directly	50	49		
13 mm resilient channels	51	49	54	49
40 mm wood furring	53	54	55	59
50 mm Z-bars	52	52	59	64
65 mm steel studs	58	57	60	72
75 mm Z-bars	57		61	

When the first layer of drywall is added (as shown in Figure 3.7), the transmission loss increases above a certain cross-over frequency, about 200 Hz, and decreases below that frequency relative to the bare blocks. The mass-air-mass resonance occurs at around 100 Hz in this case. Adding the same drywall system on the second side (as shown in Figure 3.7) improves the transmission loss further above the cross-over frequency, but below that frequency, the transmission loss gets still worse. In this case, because the air gap is too small, the effect of the mass-air mass resonance is to pull down the TL curves at frequencies within the range of the STC calculations and the STC is reduced in one case. The vertical line shows the lower limit of the STC calculation.

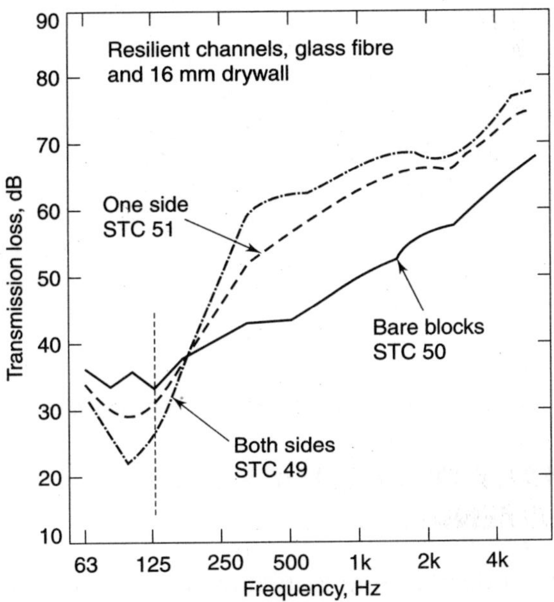

Figure 3.7 Poor transmission loss at low frequencies limits overall STC improvement suggested by better performance at higher frequencies

Despite the fact that applying treatment on both sides increases the TL at mid- and high frequencies, the STC is limited by the 8 dB rule at 125 Hz. (As mentioned above, the 8 dB rule states that no TL value can be more than 8 dB below the STC curve.) Thus the STC provides some protection, but only some, against poor TL at lower frequencies.

If the mass-air-mass resonance is low enough, these detrimental effects occur below 125 Hz (as shown in Figure 3.8) and the STC is not reduced. However, there are still reductions in the low frequency transmission loss. Changes in low frequency sound insulation may make a system unsuitable for a use where low frequency noise is expected to be a problem. Where low frequency transmission loss is important, sound transmission loss curves should be examined to be sure that any proposed wall system is good enough. These examples show why it is important to remember that STC is an average and that only data from 125 Hz upward are used in its calculation.

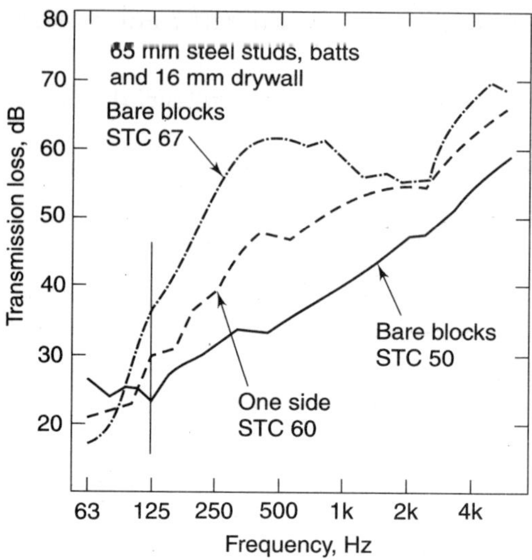

Figure 3.8 Poor transmission loss performances

3.4   SUMMARY OF FINDINGS AND RECOMMENDATIONS

As a result of the understanding gained through the measurement series, Table 3.2 gives recommended cavity depths to be used behind drywall attached to concrete blocks.

Table 3.2 Recommended Minimum Cavity Depths, mm, Behind Drywall Layers added to Concrete Block

	Number of layers of 13 mm drywall		Number of layers of 16 mm drywall	
	1	2	1	2
No sound absorbing material	90	45	75	40
With sound absorbing material	65	30	55	30

The cavity thicknesses are somewhat larger than those normally used; the recommendations ensure that the added materials do not decrease the STC of the wall system relative to the bare blocks. There are some indications that smaller cavity thicknesses may be acceptable with light weight, more porous blocks. Research is in progress to clarify this.

Essential points from this study are:

- mass-air-mass resonance has a great deal of effect on STC, and much more on low frequency sound transmission loss;

- the greater the airspace, the lower the mass-air-mass resonance and the greater the STC;
- the addition of sound absorbing material lowers mass-air-mass resonance;
- the use of resilient connections instead of rigid supports increases high frequency performance but the STC rating may be still controlled by low frequency behaviour;
- if adding a layer on one side causes a detrimental resonance, then adding a similar layer on the second side makes the resonance worse.

Figure 3.9 illustrates the differences that can be achieved by doing things correctly. For bare blocks, the STC is about 50. For the price of some sound absorbing material and a few centimetres of space, an STC of 72 can be obtained.

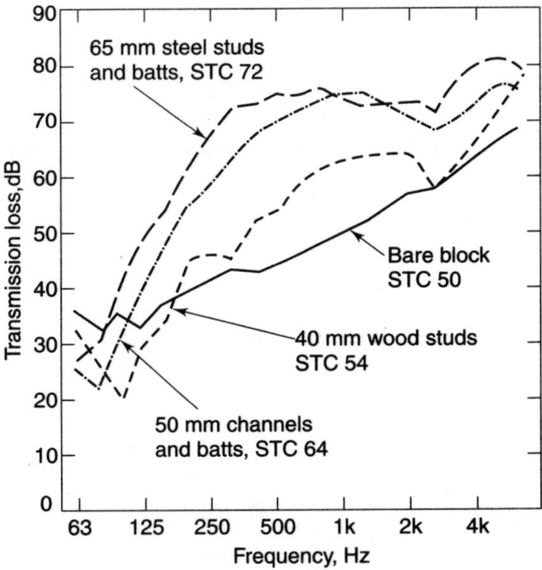

Figure 3.9 Significant improvements in transmission loss achieved with simple constructions

3.5   PLUMBING NOISE

Many articles give general advice on how to deal with plumbing noise. The most frequent recommendation is to mount the pipes and all devices resiliently (as shown in Figure 3.10). Many questions about the effectiveness of these techniques in Indian construction have, until now, remained unanswered. Recently NPL collaborated in a study of some resilient mounting techniques and other methods that might be used to control plumbing noise in buildings.[7]

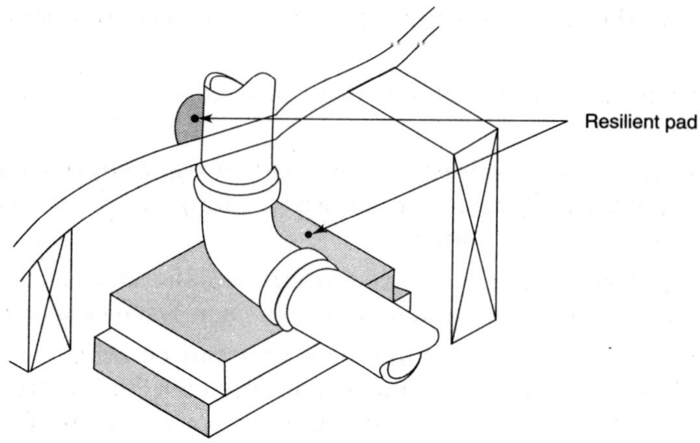

Figure 3.10 Possible means of isolating plumbing from sound-conducting elements

The device used to generate the noise in the plumbing was a standard source that is used tests of plumbing noise (as shown in Figure 3.11). Water is forced to pass through two obstructions in the pipe, the first with four small holes, the second with one. This creates a lot of turbulence and noise; about 5 dB more noise than a conventional.

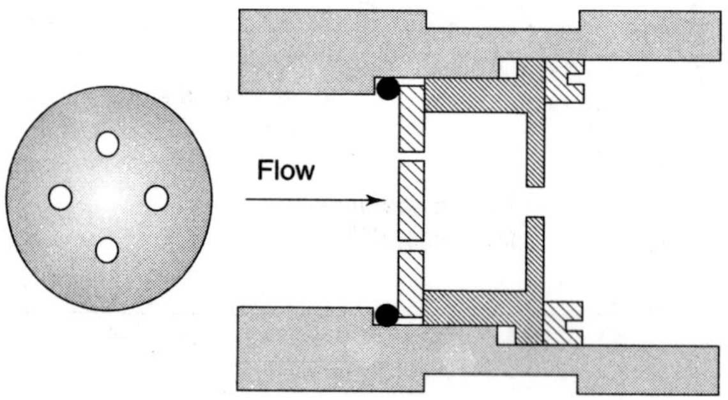

Figure 3.11 Cross section through the standard noise source

3.6   RESILIENT WRAPPINGS

Adding resilient wrapping (as shown in Figure 3.12) between the pipe and the clamping hardware is one method of achieving a resilient mount and means that vibrations in the wall of the pipe are not so easily transmitted to the wood studs and thence to the drywall. The objective is to interrupt the path the sound must follow on its way to being radiated.

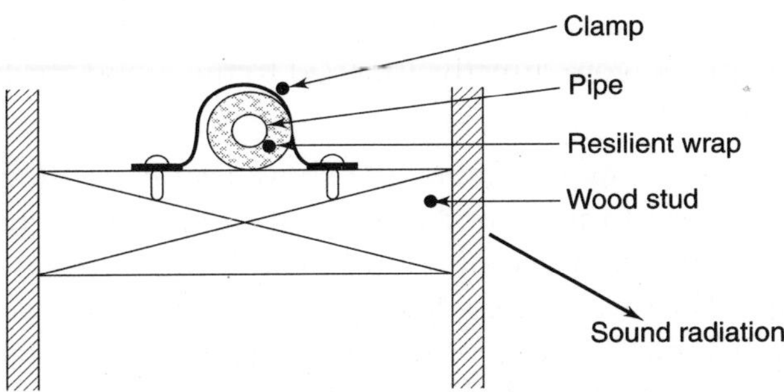

Figure 3.12 Plumbing noise test set-up

Table 3.3 shows the advantages of using different resilient wrappings around a copper pipe. A soft material can reduce noise by about 20 dBA relative to the rigid clamps. Generally the softer the material, the greater the noise reduction. Notice that the noise level is reduced; it is not eliminated. Adding wedges simulates errors. The wedges increased the noise levels by about 8 to 10 dBA.

Table 3.3 A-weighted noise levels measured with various attachments of pipes to studs

Resilient material	Measured Noise Levels, dBA Pipe diameter		
Rigid clamps	73	71	72
2 Mm cork + clamp	68	64	63
13 Mm felt + clamp	64	59	56
Solid neoprene + clamp	59	58	57
Neoprene foam + clamp	54	54	50
No clamps	47		
Neoprene foam + 1 wedge	62		
Neoprene foam + 2 wedges	65		
Neoprene foam + 3 wedges	65		

3.7   WALL SYSTEM MODIFICATIONS

As an alternative to resilient mounting of pipes, or where noise reduction is required in an existing installation, one might consider changes to the wall system. Several means of improving the basic wall were investigated. In all cases the pipes were directly attached to the wood studs; no resilient materials were used.

The results (as shown in Table 3.4) show that even though the pipes are directly clamped to the wood studs, substantial noise reductions can be achieved through the use of sound absorbing material and resilient metal channels. The lowest noise level given in Table 3.4 is about the same as that given in Table 3.3, (except for the measurement in Table 3.3 where the absence of clamps provided for no contact at all with the studs). It is tempting in problem situations to blow sound absorbing material, either glass or cellulose fibre, into the wall. This table shows that both materials give about the same noise levels and that better results are obtained by introducing resilient metal channels to support the drywall.

Table 3.4 Noise Levels Measured from Modified Wall Systems

Wall Finish	Measured Noise Levels, dBA Pipe diameter	
	13 mm	25 mm
1 layer drywall	73	71
1 Layer drywall with batts in cavity	73	68
2 layers drywall	70	66
1 layer drywall with cellulose fibre in cavity	--	67
2 layers drywall with batts in cavity	68	66
1 layer drywall on resilient channels with batts in cavity	64	62
2 layers drywall on resilient channels with batts in cavity	56	56

3.8 COMBINED APPROACH

Using resiliently mounted pipes and improving the wall system gives even greater plumbing noise reduction. Table 5 gives results for several types of wall where the pipes were supported using 13 mm thick neoprene foam resilient wrapping. The best construction in this case is about 30 dBA quieter than the wall with a single layer of drywall and the pipes solidly mounted.

Table 3.5 Noise Levels with Wall Modification and 13 mm Thick Neoprene Foam Resilient Mounting

Wall finish	Measured Noise Levels, dBA Pipe diameter	
1 layer drywall	54	55
1 Layer drywall with batts in cavity	51	50
2 layers drywall	51	51
2 layers drywall with batts in cavity	48	47
1 layer drywall on resilient channels with batts in cavity	44	44
2 layers drywall on resilient channels with batts in cavity	42	4

3.9   COMPARISON OF PIPE MATERIALS

Different pipe materials may be expected to transmit sound energy differently. Measurements were made with two commonly available materials used for supply pipes, copper and plastic. The plastic pipe was Schedule 80 pipe with a wall thickness of 4 to 5 mm depending on diameter. Comparisons are given in Table 6 for the average of 13, 19 and 25 mm diameters of these two types of pipe, with and without a resilient wrapping. The plastic pipes are significantly quieter than the copper pipes when no resilient wrapping is used but when the soft foam wrap is used, there is little difference. However, if there is unintended contact, the noise generated will be less with the plastic pipe.

Table 3.6 Comparison of Sound Transmitted by Copper and Plastic Pipe

Material	Measured Noise Levels	
	Solid Clamps	Neoprene foam + Clamps
Copper	72	54
Plastic	62	52

3.10 CONCLUSIONS

The general conclusions to be drawn from this study are that the use of resilient supports for plumbing pipes and other systems is very important. Noise reductions up to about 15 dBA can be obtained relative to systems where no resilient mounts are used for pipes.

Adding extra drywall always reduces the noise; adding resilient metal channels is more effective and provides some margin if construction errors result in accidental solid contact between pipes and structure.

3.11 FLANKING NOISE

In the laboratory we take great care to mount specimens so that the only path for sound is through the specimen; there is no solid connection between the specimen and the rooms on either side (as shown in Figure 3.13). This can be done in the laboratory, but not so easily in a building. Figure 3.14 shows the many paths that sound can follow when it travels between two rooms. Ideally, one would resiliently mount all surfaces in a room to attenuate all direct and flanking paths.

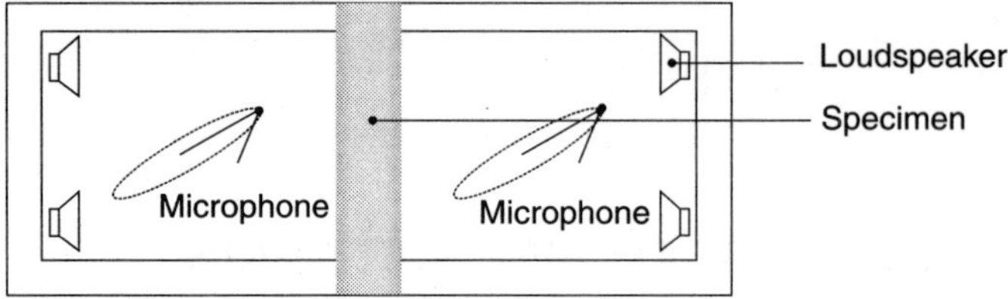

Figure 3.13 Laboratory set-ups for measurement of sound transmission loss

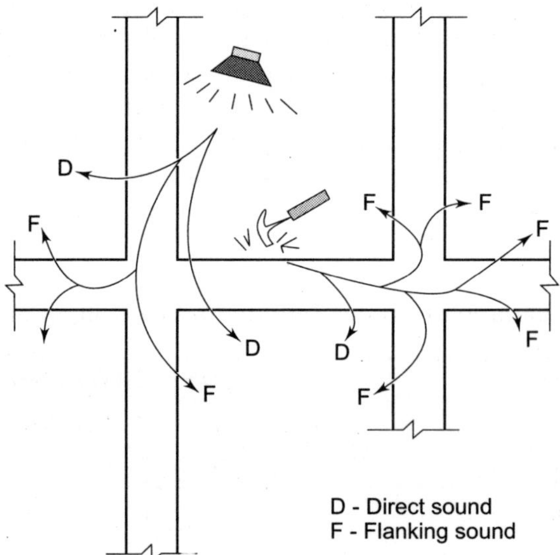

Figure 3.14 Flanking sound through building construction

A saw cut in the floor eliminates a horizontal flanking path along the plywood (as shown in Figure 3.15). Such saw cuts are recommended. But there is also the vertical path to be considered. Sound, especially impact noise, can travel down the walls to the space below (as shown in Figure 3.16). To reduce transmission along this path, one can use resilient metal channels (as shown in Figure 3.17). The channels have the advantage that they also help to reduce plumbing noise. This approaches the ideal situation in building noise control, where all surfaces in a room to be protected are mounted resiliently.

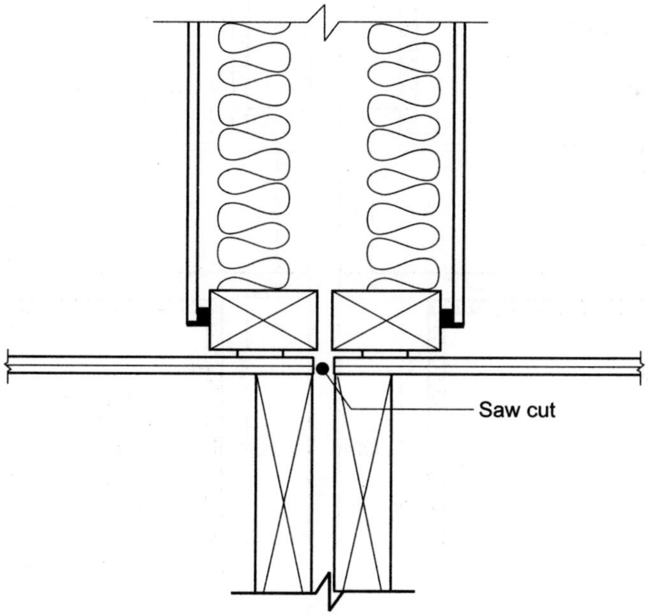

Figure 3.15 Saw cut to break sound flanking through floor

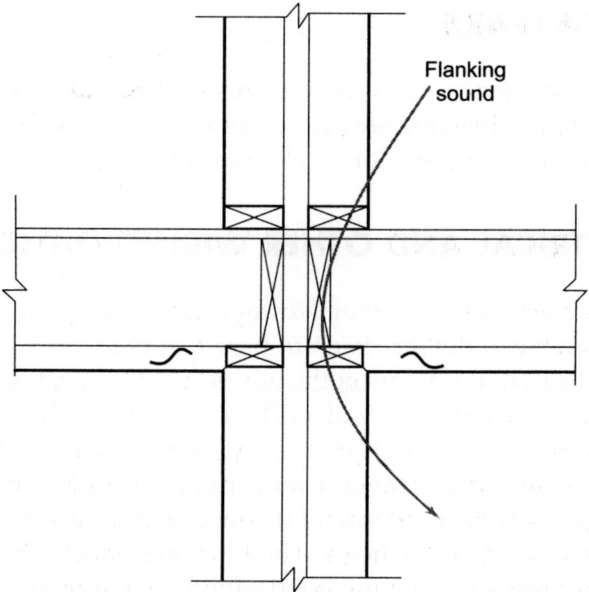

Figure 3.16 Sound flanking vertically through walls

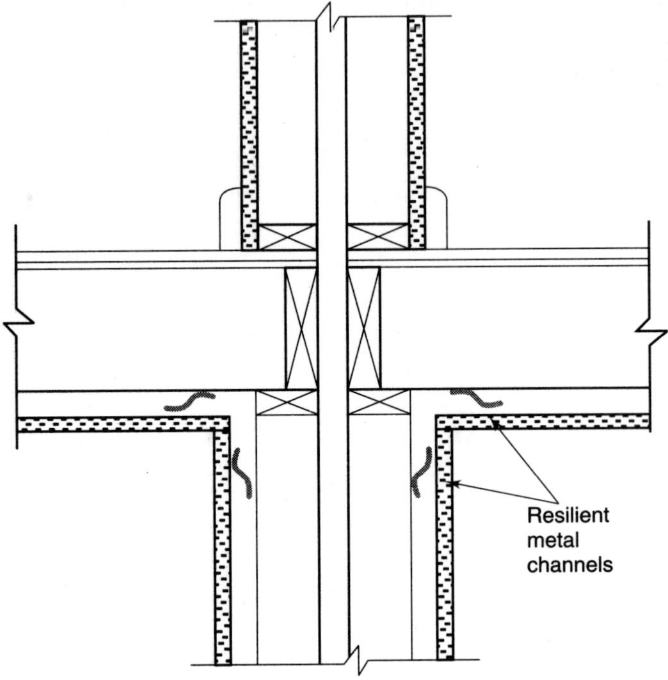

Figure 3.17 Resilient channels used to reduce flanking noise

3.12 NOISE LEAKS

Two types of noise leaks are common causes of sound problems in small buildings; leakage through electrical outlets is relatively minor, while leakage around interior partitions can be major.

3.13 ELECTRICAL AND OTHER WIRING OUTLETS

Leaks around electrical and other wiring outlets are a common problem, especially when the outlets are back-to-back (as shown in Figure 3.18). One recommended solution is to offset the boxes (as shown in Figure 3.19). This is especially effective when the wall is filled with sound absorbing material. Leaks usually permit only high-frequency sound to pass through. Forcing high-frequency sound to travel a long path through sound absorbing material is a good way to attenuate it, since sound absorbing material is most effective at high frequencies. The blocking panels shown in Figure 3.20 are another way of achieving good sound insulation while still having back-to-back outlets.

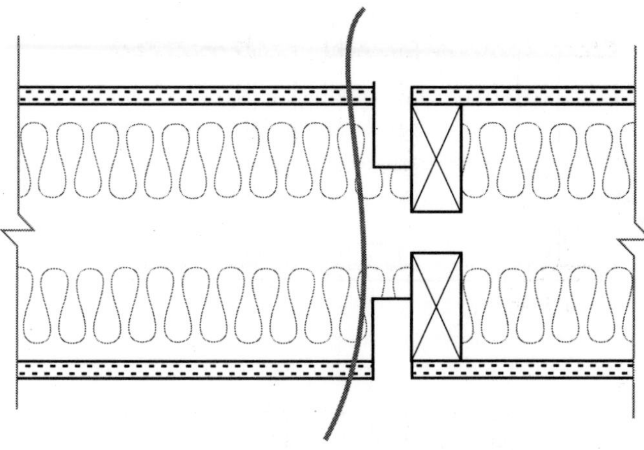

Figure 3.18 Leakage through electrical boxes

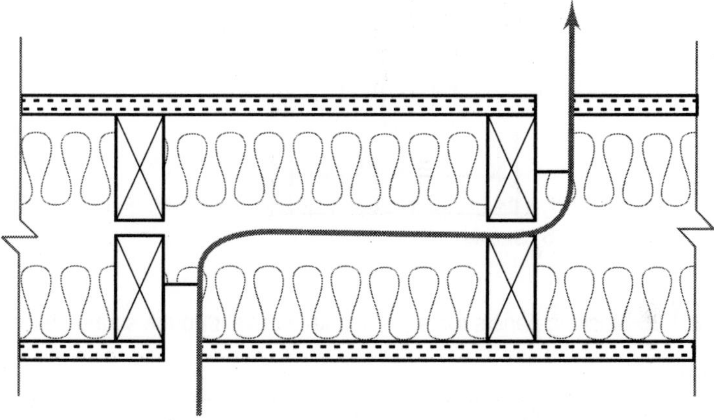

Figure 3.19 Offset boxes make the sound path longer

3.14 PARTITION AND CEILING SPACES

This major leak is common in office buildings with suspended ceilings. Sound is transmitted via the space above the ceiling where the common wall does not extend to the slab above (as shown in Figure 3.21). There are two approaches to dealing with this problem: either the partition is made full height or the attenuation of the path through the plenum and ceiling is increased. For each of these approaches there are variants.

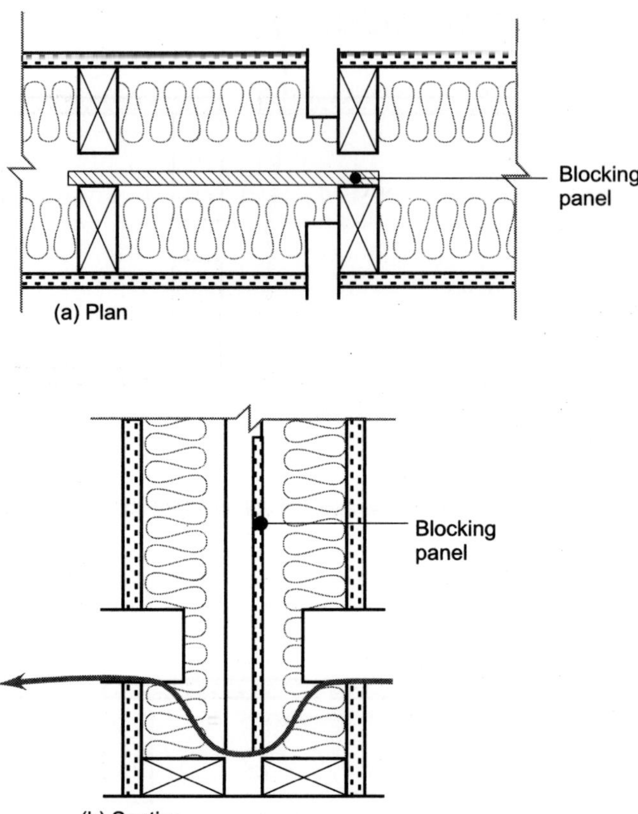

Figure 3.20 Blocking panels force the sound to travel through sound absorbing material

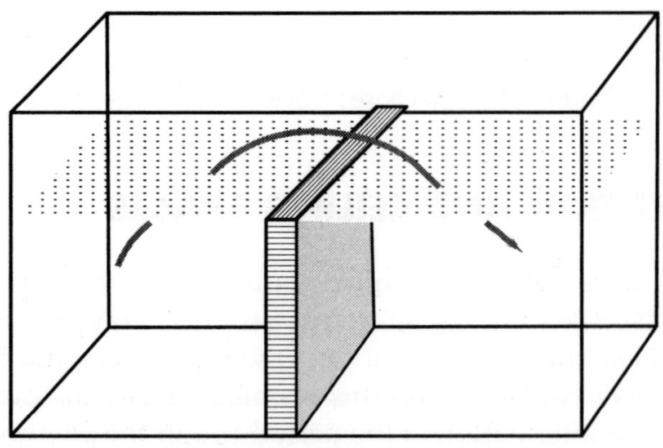

Figure 3.21 Noise leak through plenum space

3.15 CEILING TREATMENT

Three approaches for increasing the plenum/ceiling attenuation were tested[6] (as shown in Figure 3.22). In the first case, a layer of 6 mm drywall was laid on top of tiles. The test 4 results are presented in terms of increased noise isolation class (NIC). NIC is a measure of sound insulation that is very similar to sound transmission class (STC). However, it includes the effects of the room, whereas these are removed by calculation before the STC is worked out. Adding 6 mm drywall increased the sound insulation by 5 dB, to NIC 37.

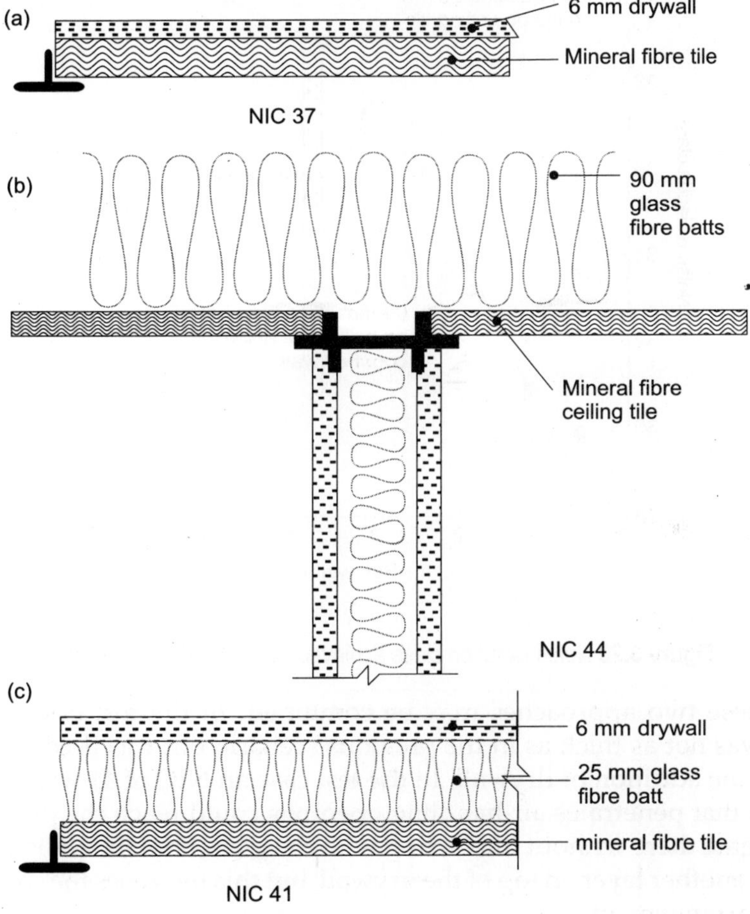

Figure 3.22 Plenum/ceiling treatments

(a) 6 mm drywall on mineral fibre tile

(b) 90 mm glass fibre batts on ceiling tiles

(c) 25 mm glass fibre batts between the drywall and the tiles

Adding 90 mm glass fibre batts on top of mineral fibre tiles is intended to absorb the sound as it propagates in the plenum. The NIC increased as the width of the batts above the wall was increased. The ceiling had typical openings for air handling. Once the width of the batts had reached about 3.5 m, there was a marked reduction in improvement when the width was further increased (as shown in Figure 3.23). This result is probably specific to the particular test arrangement.

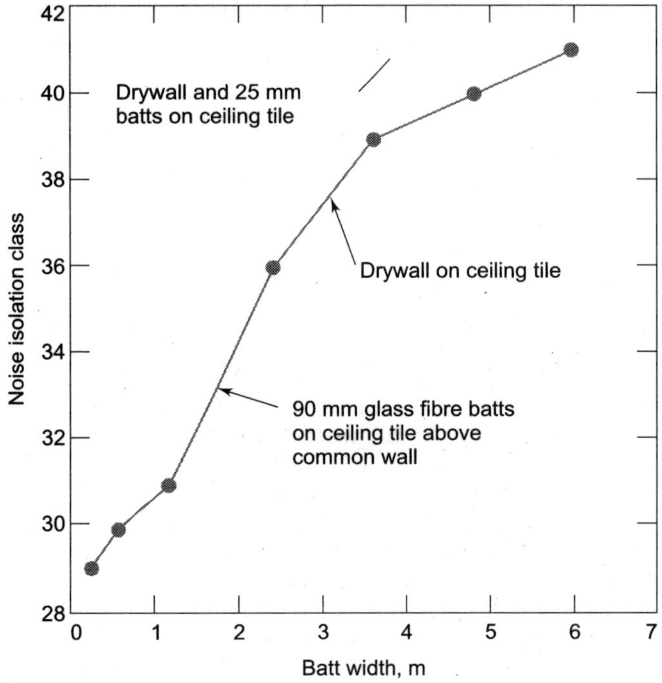

Figure 3.23 Noise isolation class achieved by plenum/ceiling treatment

These two approaches may be combined. In the test case the glass fibre was not as thick as in the previous scenario but this is compensated for by the addition of drywall. In the test case an NIC of 41 was achieved. Sound that penetrates air-handling openings could reach the plenum and propagate there without interacting with the glass fibre. It would be better to add another layer on top of the drywall, but this increases the complexity of the arrangement.

Partition Extension

The gap between the top of the wall and the floor slab above may be blocked in several ways. Simply extending the wall should give sound insulation values appropriate for the wall (as shown in Figure 3.24). The ceiling boards already provide some attenuation, so the blocking panels need not provide as much attenuation as the wall itself (as shown in Figure 3.25).

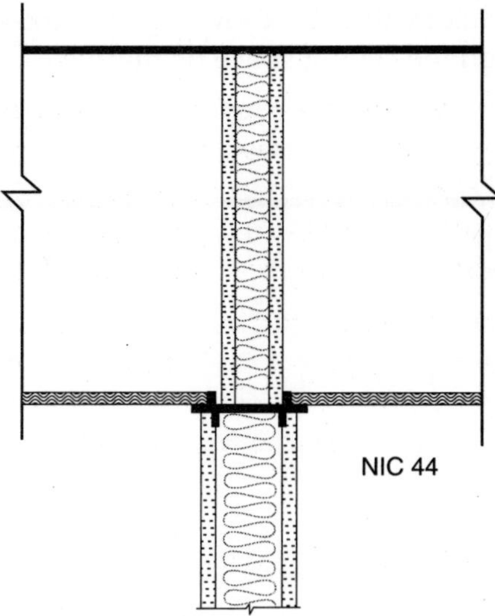

Figure 3.24 Wall construction extended through plenum space

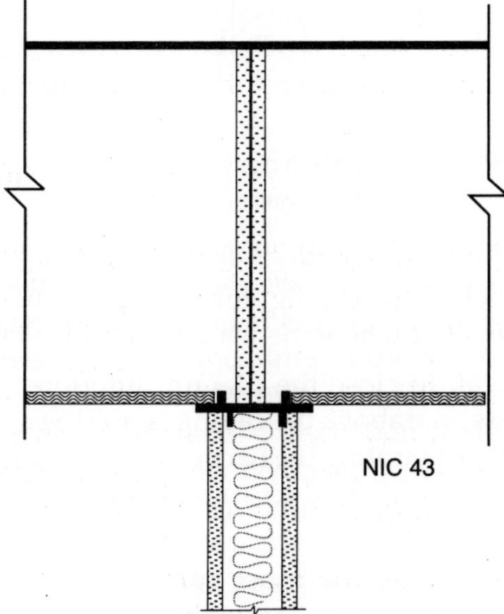

Figure 3.25 Blocking panels carried through plenum space

Adding panels to extend the wall can be very difficult because of wires, pipes and other services in the plenum space. Another technique that works well is to fill the space above the wall with batts of glass fibre or mineral wool (as shown in Figure 3.26). This is easy to pack in around pipes and other obstructions. This treatment, known as the fuzz-wall approach, can make the path through the ceiling space negligible.

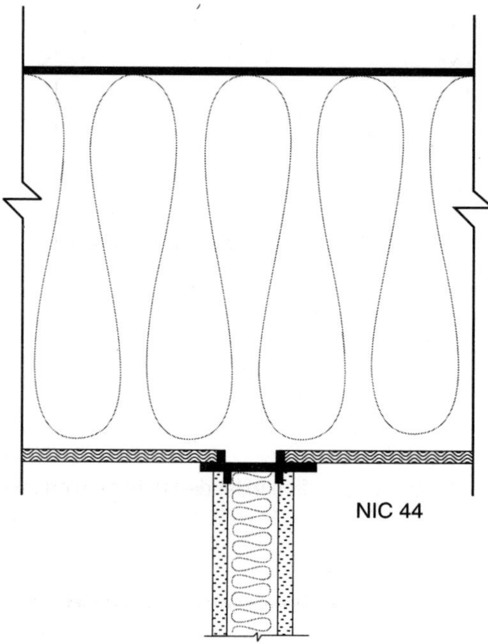

Figure 3.26 Plenum packed with glass fibre above the partition wall -
the fuzz wall approach

There is little point in attenuating sound in the plenum if there are other major sound leaks below the ceiling, for example, where the ceiling tiles meet the top of the wall or where the wall meets side walls or the floor.

Extending the wall to close the plenum interferes with airflow and ventilation when the space above the ceiling is used as a return air plenum. Additional ductwork penetrating the barrier will be necessary to restore the airflow. Since this ductwork introduces a path for sound, it should be lined with sound absorbing material.

Sound leaks are one of the most common causes of poor sound insulation. Repairing leaks or preventing their occurrence is not difficult in most. Often caulking is all that is necessary. The cases examined here are a little more complicated but they too can be controlled using sound barrier materials and sound absorbing materials.

3.16 SUMMARY

Noise control in buildings usually involves no more than the correct application of solid barrier materials, resilient materials and sound absorbing materials. The basic principles are fairly simple and complications such as the mass-air-mass resonance mentioned above are relatively infrequent.

4

Sound Transmission Through Building Elements

4.1 INTRODUCTION

If significant noise reduction is required between two rooms, the wall or floor separating them must transmit only a tiny fraction of the sound energy striking it. The ratio of the sound energy striking the wall to the transmitted sound energy, expressed in decibels, is called the transmission loss (TL). The less sound energy is transmitted the higher the transmission loss. For common applications, transmission loss ranging from 30 to 70 dB may be required -these correspond to transmission of from 0.1% to 0.00001% of the sound energy. Such reductions cannot be achieved unless cracks and holes are carefully sealed -- if air can pass from one room to the other, then sound will also be transmitted. The discussions presented in this paper (except the discussion of doors) assume the elimination of air leaks.

For typical partitions, the transmission loss is much smaller for low frequency sounds than for high frequency sounds. For convenient comparison of different constructions, it is necessary to have a single rating that puts the performance at all frequencies into perspective. The sound transmission class (STC) is most commonly used. For sounds like human speech, the STC provides an accurate rating of perceived noise reduction. However, when the noise has strong low frequency content (noise from ventilating fans or music with a strong bass line), the effective noise reduction may be appreciably less than the STC would suggest. Despite this shortcoming, the STC is used as the primary rating throughout this discussion because of its widespread use by designers. The additional problem of low frequency noise is addressed frequently.

This Chapter begins with a discussion of the basic factors that control sound transmission through any building component. An understanding of the basic principles can lead to design economies and the avoidance of error. Small changes in the arrangement of materials can yield large changes in noise reduction with little or no increase in cost. Later sections examine the special features that affect the noise reduction performance of typical walls, windows, doors and floors.

Mass Law

The first variable normally considered in predicting sound transmission through a panel is its mass per unit area. An increase in transmission loss is expected with increasing mass, because the heavier the panel, the less it vibrates in response to sound waves, hence the less sound energy it radiates on the other side. The mass law is a theoretical expression that applies to thin panels of most common building materials within a limited frequency range. The mass law predicts that TL will increase by approximately 6 dB for each doubling of the surface mass or the frequency (as shown in Figure 4.1). An increase in mass can be achieved either by increasing the material's thickness or by selecting denser material. The unit surface mass for 1 mm thicknesses of some common building materials is given in Table 4.1. These values should be multiplied by the total thickness of the material to get its overall surface mass. The transmission loss at a given frequency can then be found from Figure 4.1 or by using the following equation:

$$TL = 20 \log (mf) - 48 \qquad \qquad ...(4.1)$$

where:

TL is transmission loss (dB), m is surface mass (kg/m^2), f is frequency (Hz).

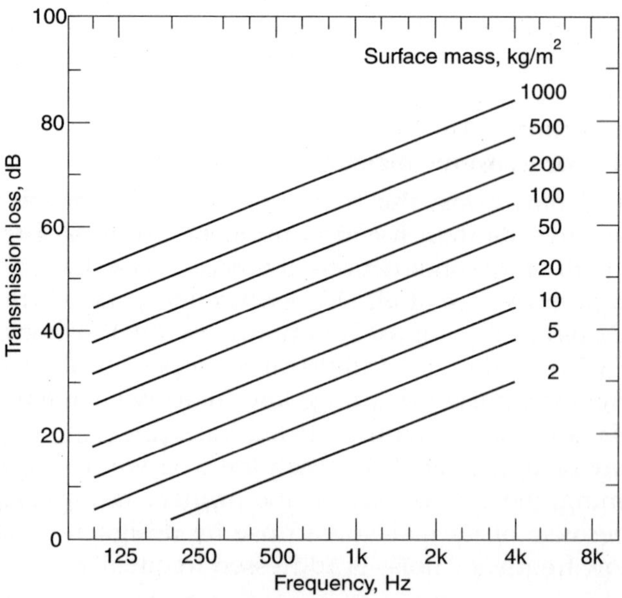

Figure 4.1

Table 4.1 Surface Mass for 1 mm Thicknesses and Constant a (for Calculation of Coincidence Frequency) for Some Common Building Materials.

Material	Surface Mass kg/m^2 per mm thickness	Constant A Hz · mm
Aluminum	2.7	12,900
Concrete, dense poured	2.3	18,700
Hollow concrete block	1.1	20,900
Fir timber	0.55	8,900
Glass	2.5	15,200
Lead	11.0	55,000
Plexiglas or Lucite	1.15	30,800
Steel	7.7	12,700
Gypsum board	0.8	39,000
Plywood	0.6	21,700

In principle all noise transmission problems could be solved by increasing the partition mass; however, practical considerations limit the usefulness of this option.

Coincidence Dip

The transmission loss measured in practice often differs from that determined theoretically from the mass law. The typical TL for some common materials is illustrated in Figure 4.2. As one would expect from its greater mass, the concrete slab exhibits much higher TL than a sheet of gypsum board or plywood. However, significant deviations from the mass law are evident for all three materials.

The curves for gypsum board and plywood (both about 10 kg/m^2) have very similar transmission loss at low frequencies, increasing steadily, in good agreement with the mass law. At higher frequencies, each exhibits an appreciable dip, called the coincidence dip. This dip is centered at the coincidence frequency, which depends on the material's stiffness and its thickness. The coincidence dip of the plywood occurs at a lower frequency than that for gypsum because the plywood is stiffer. Even though the plywood and gypsum board weigh the same, the plywood has a lower STC because of this dip.

The coincidence frequency for building materials listed in Table 4.1 can be calculated by dividing the appropriate value of the constant A by the thickness in mm. For example, concrete has A = 18,700, so a layer of concrete 100 mm thick would have a coincidence frequency near 190 Hz. The TL data (ad shown in Figure 4.2) fall below the mass law for all frequencies above the coincidence frequency.

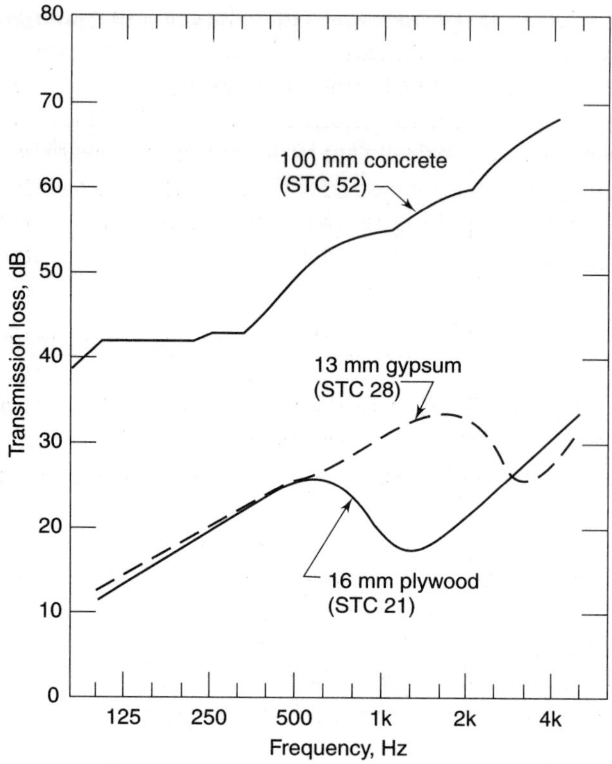

Figure 4.2

Concrete and plywood in common thicknesses are strongly affected by coincidence because the coincidence frequency tends to fall in the range important for building acoustics (80 to 4000 Hz). For a flexible panel such as thin sheet metal, the steady increase in transmission loss continues throughout the frequency range, because the coincidence dip is far above 4000 Hz. 'Limp' materials, such as thin sheets of lead or plastic, do not exhibit an appreciable coincidence dip. Fortunately, significant problems due to coincidence can normally be avoided by proper use of less expensive materials.

The depth and width of the coincidence dip are determined by losses of sound energy in the material, and energy transfer to the supporting structure. The greater these losses, the shallower and broader the coincidence dip, and the less it affects the STC. When two layers of material such as gypsum board are glued firmly together, they behave like a single thick layer, with an associated lowering of the coincidence frequency. However, if the layers are attached more loosely (with screws for example), permitting surface slippage during bending motions, the coincidence dip does not shift to a lower frequency and friction between layers can increase the energy losses, giving higher TL near coincidence. For the same reason, laminated glass with thin layers of suitable plastic between the layers of glass can provide much higher STC than solid glass of the same thickness.

Cavity Constructions

Partitions divided into two or more layers by intervening cavities can provide high STC values with lightweight construction. However, this potential increase in TL may be limited by structural coupling between the layers; vibration transfer through the structure is the acoustical equivalent of an electrical short circuit. The transmission loss of a cavity construction depends strongly on the characteristics of the intervening space.

An ideal cavity partition would have no structural connection between the layers. Sound striking the partition would cause vibration of the exposed layer. This layer would transmit some of its vibrational energy to the air in the cavity, which would transmit some of its vibrational energy to the second layer, which in turn would radiate some of its vibrational energy into the room on the other side. Because only a fraction of the sound energy is transmitted at each of these steps, the cavity partition provides increased noise reduction.

In general, the larger the space between the two layers, the higher the transmission loss. This is illustrated by the data in Figure 4.3 for two layers of glass with different spacings. The larger spacing gives higher transmission loss throughout the frequency range. Very similar transmission loss would be observed for other materials such as gypsum board with the same spacings. Because increasing the space between the two surfaces gives higher transmission loss, it is preferable to use as large a spacing as possible.

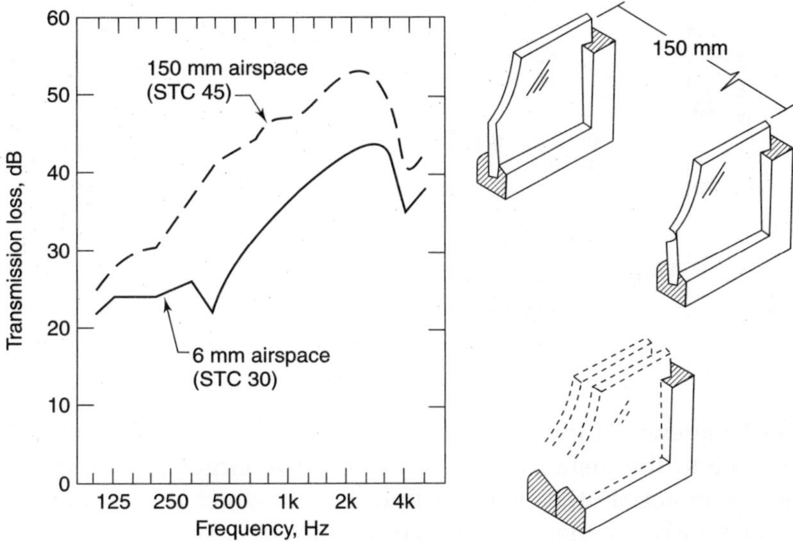

Figure 4.3

Mass-Air-mass Resonance

There is a practical basis for setting a minimum spacing between the layers. The lower curve in Figure 4.3 (for 6 mm spacing) has a sharp dip at frequencies just below 500 Hz. This is due to the mass-air-mass resonance. The air trapped in the cavity between the layers acts as a spring transferring vibrational energy from one layer to the other. This energy transfer is significant only in a small frequency range, where it causes a sharp lowering of the TL.

The effect of this resonance is illustrated in Figure 4.4, which compares the transmission loss of a single layer of 3 mm glass with the transmission loss given in Figure 4.3 for double glass with 6 mm air space. Near the resonance frequency (250 to 500 Hz) the TL of the double glass is actually less than that for a single layer of the same glass. At lower frequencies, the double glass has higher transmission loss; this is due to doubling the mass of glass, and is essentially the same result that would be obtained if the two layers of glass were glued together. Only at frequencies appreciably above the resonance does the airspace provide significant acoustical benefit. (The dips that occur at about 4000 Hz are due to coincidence, not the mass-air-mass resonance).

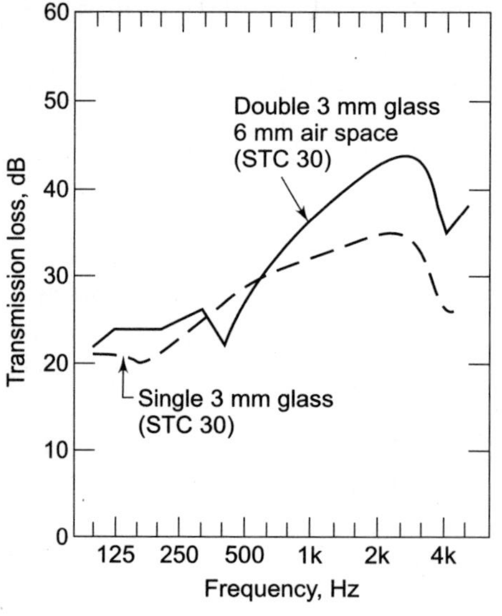

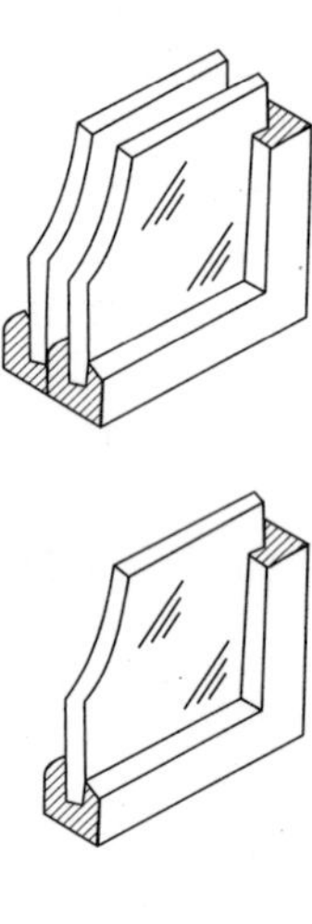

Figure 4.4

The frequency of the mass-air-mass resonance depends on the mass of the layers and the distance between them. The larger the air space or the heavier the materials, the lower the frequency at which the resonance occurs. The frequency of the mass-air-mass resonance (f_{mam}) can be calculated from the equation:

$$f_{mam} = 1900\sqrt{(m_1 + m_1)/dm_1m_2} \qquad\qquad ...(4.2)$$

where:

f_{mam} is the mass-air-mass resonance frequency, m_1 is the surface mass of the first layer (kg/m^2), m_2 is the surface mass of the second layer (kg/m^2), d is their separation (mm).

To maximize the improvement due to the airspace, partitions should be designed so that the mass-air-mass resonance is at as low a frequency as possible. As a general guideline, designing for a mass-air-mass resonance of 80 Hz makes optimum use of wall or floor materials. Many common partition designs do not meet this criterion. The resulting deficiencies in their low frequency TL can be readily demonstrated by a stereo system in the adjacent room; the melody line is barely audible, but the bass line thumps through.

Filling the cavity with absorptive material can increase the transmission loss substantially, especially when the cavity is large. This is obviously impractical for windows because the absorptive batts are not transparent, but it is a common way to reduce sound transmission through walls or floors. The density of the sound-absorbing material is not very important and using batts thicker than two-thirds of the cavity depth provides little additional increase in STC. Materials used for thermal insulation, such as cellulose fibre, glass fibre or mineral wool insulation, are well suited for this purpose.

Adding absorptive material to the cavity is beneficial only if structural connections between the surfaces do not transmit much vibrational energy. For example, adding glass fibre between the studs of a common wall (wood stud wall with gypsum board attached directly to both faces of the studs) has little effect because it does not alter the dominant vibration transmission path through the studs. Figure 4.5 compares the transmission loss of a conventional wood stud wall with that when both layers of gypsum board are attached to the same side of the studs, eliminating the airspace. Because of the vibration transmission through the studs, the airspace between the two layers of gypsum board gives little improvement. At low frequencies the cavity actually gives a lower TL because of the mass- air-mass resonance.

4.2 SPECIFIC WALL DESIGNS

For gypsum wallboard construction, there are several practical ways of reducing mechanical connections between partition layers; these include staggered wood studs, separate rows of wood or steel studs or, with a single row of wood studs, resilient metal channels to support the wallboard layers independency of each other. Non-load-bearing steel studs (typically made from 24 gauge sheet steel) are usually resilient enough to provide adequate mechanical decoupling between layers of gypsum board applied to both sides. For load-bearing steel studs, good results have also been obtained through the use of resilient channels. Some of these constructions are illustrated in Figure 4.6.

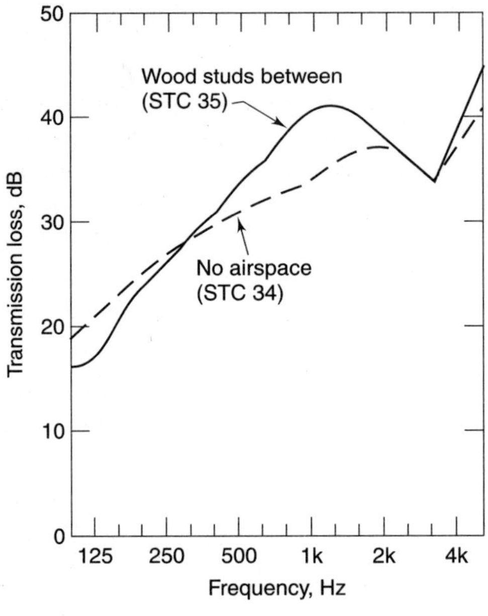

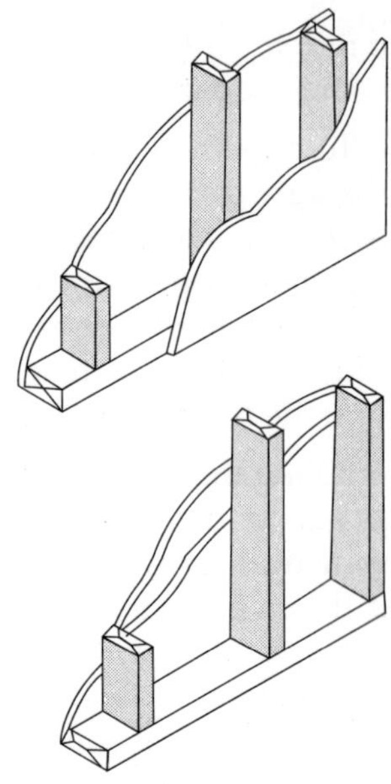

Figure 4.5

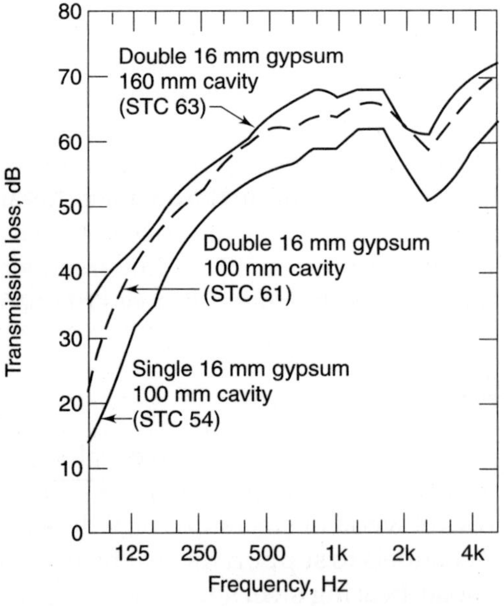

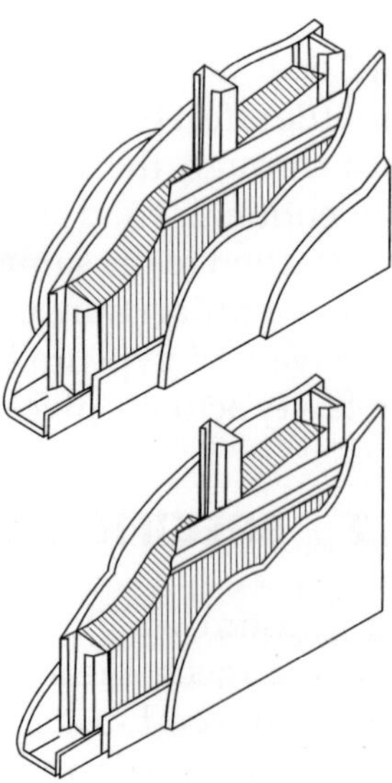

Figure 4.6

Table 4.2 gives approximate STC values for some gypsum board walls. Some deviations from the listed values should be expected because of variations in type of wallboard (e.g., fire-rated versus standard), wallboard attachment, type and thickness of absorptive material, and stiffness of steel studs or resilient channels. Other constructions with minimal structural connection and similar cavity depth would give similar transmission losses. Where two layers of gypsum board are indicated in Table 4.2, the layers are assumed to be screwed together. If resilient channels were used between two layers of gypsum board, the STC would be the same or lower than that for the same construction with a single layer, because of the mass-air-mass resonance for this small airspace.

In contrast to the values listed in Table 4.2, the common internal partition used in single family homes (drywall attached directly to both sides of wood studs) has an STC rating of about 34 (as shown in Figure 4.5). The addition of sound- absorbing material to the cavity would increase the STC by less than 5, because most of the sound energy is transmitted through the studs. A greater improvement in the STC could be obtained by reducing rigid mechanical connections between the layers.

Table 4.2 Approximate STC Ratings for Walls with 13 mm Gypsum Board on both Surfaces

	No cavity absorption Layers of gypsum board			With absorption in cavity Layers of gypsum board		
Structural Support	1+1	1+2	2+2	1+1	1+2	2+2
90 mm 24 gauge steel studs	39	44	50	45	49	53
38 x 89 mm wood studs with resilient steel	40	45	51	48	52	54
channels on one side on two sides	40	46	52	49	52	55
Staggered 38 x 89 mm wood studs	40	46	52	49	52	55
150 mm load-bearing steel studs with resilient metal channels on one side	45	51	56	56	58	61
Double 38 x 89 mm wood studs with 25 mm gap between	46	52	57	57	60	63

A partition with STC greater than 50 will provide speech privacy, but may not provide satisfactory isolation against noise with a strong low frequency component. In Figure 4.6 the bottom curve is the transmission loss for a partition with a single layer of 16 mm gypsum board on each side, supported by 90 mm structural steel studs with 10 mm resilient channels at one side; the STC is 54. At the lowest frequencies, the transmission loss drops rapidly because of the mass-air-mass resonance; hence bass notes will be transmitted quite well.

To improve the transmission loss, it is necessary to increase the weight of the surface layers and/or increase the distance between the surfaces. Adding a second layer of the same type of gypsum wallboard on each face increases the STC to 61 (as shown in Figure 4.6), and the sharp drop in transmission loss near 100 Hz is shifted to lower frequency. Increasing the cavity space (by replacing the 90 mm studs with 150 mm studs) gives an even higher STC, and greatly improves the TL at the lowest frequencies. For the mid and high frequencies, vibration transmission through the studs limits the improvement due to the increased space.

Figure 4.7 compares the performance of the best case from Figure 6 with the TL for a double wood stud partition. The slightly larger cavity depth and reduced structural connection through the double wood stud system give the same STC with lighter surfaces (the mass of conventional 12 mm gypsum board is approximately 30% lower than that of 16 mm gypsum board). Using the heavier gypsum board with the double wood studs would give even better performance.

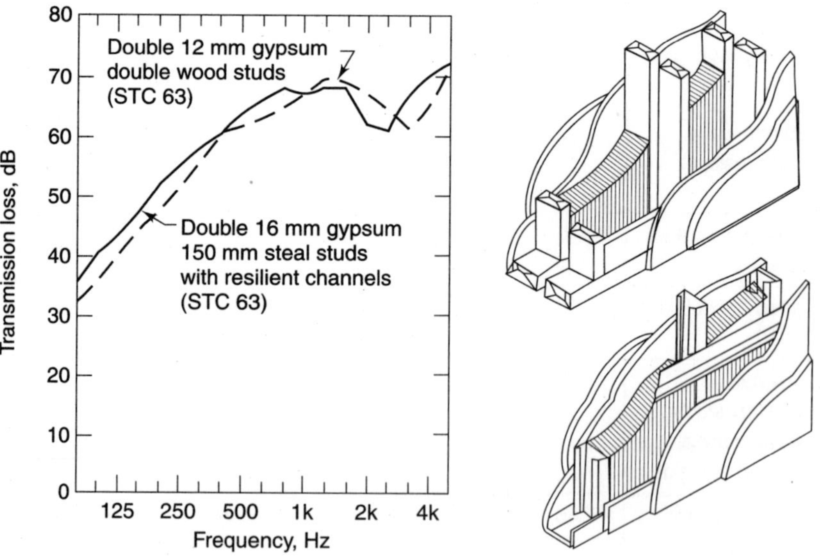

Figure 4.7

In designing a wall, the importance of a large airspace should be remembered. Although adding weight will generally increase the transmission loss of a wall, adding a layer in the wrong place can reduce the effectiveness of the airspace and thus lower the transmission loss. For example, adding a layer of gypsum board to the inner face of one row of studs in the middle of the double stud wall shown in Figure 4.7 would lower the STC to about 55. Adding a layer of gypsum board to the inner face of both sets of studs would reduce the STC to below 45.

Both high STC and good low frequency performance can be achieved with masonry walls, if correctly designed and constructed (as shown in Figure 4.8). Concrete blocks, or precast or cast-in-place concrete of the same weight give similar performance. In addition to elimination of any holes through the concrete or the mortar, controlling the mass-air-mass resonance may be an important factor in obtaining good performance from these walls. Concrete block walls commonly have wallboard applied to each face as a finishing material; if the air gap behind the wallboard is too small, then the transmission loss can be reduced relative to the unfinished wall. Reduced low frequency TL caused by the mass-air-mass resonance often lowers the STC of otherwise good concrete block walls. If the wallboard is resiliently mounted (increasing the airspace and reducing rigid connections as in Figure 4.8), and absorptive material is added between the concrete and the wallboard, the transmission loss will be increased and STC values over 60 can be obtained.

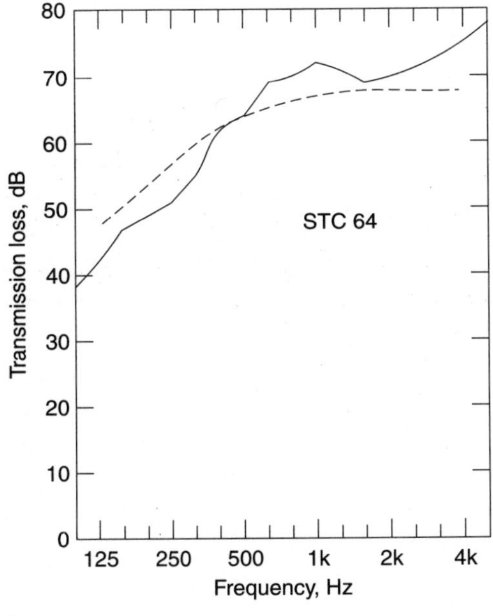

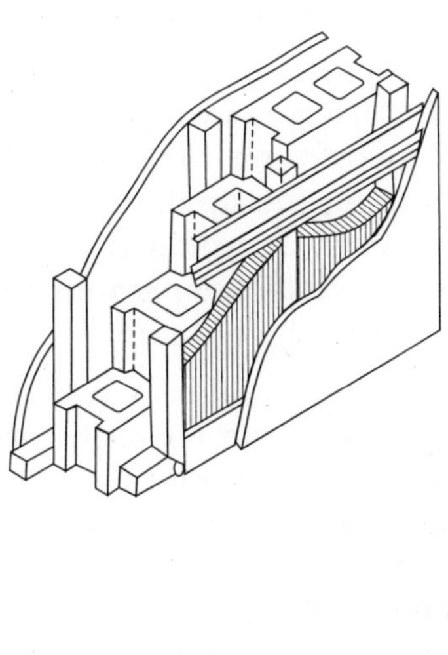

Figure 4.8

To ensure that the mass-air-mass resonance is below 80 Hz, the airspace between a single layer of wallboard and the concrete block should be at least 60 mm; for a double layer of wallboard, the space may be as small as 35 mm. These airspaces are larger than those typically used in concrete block walls, where thin wood furring strips are commonly used to attach the wallboard. Attaching the wallboard with glue also creates a small airspace; the resulting resonance lowers TL in the 200 to 400 Hz range. Porosity of some concrete blocks may increase the effective thickness of the air layer behind the wallboard, causing the mass-air-mass resonance to occur at a lower frequency than would be expected from the apparent physical dimensions. However, it is not a good idea to rely on this effect unless measurements using identical materials have shown that the chosen design will be adequate.

4.3   WINDOWS

Sound transmission through windows is governed by the same physical principles that affect walls. Practical noise control measures are influenced by the properties of the glass and the characteristics of the window assemblies.

Increasing the thickness of glass gives greater noise reduction at some frequencies but the coincidence dip, due to the stiffness of the glass, limits the improvement. For 2 mm glass, the coincidence dip is near 5000 Hz; for 18 mm glass, the dip is near 500 Hz. This shift of the coincidence dip into the mid frequencies results in only a slight increase of the STC with increasing thickness of solid glass. If an STC rating much above 30 is required, a single layer of solid glass is not a practical choice.

Laminated glass (two or more layers of glass bonded together by thin plastic inter-layers) can provide much higher TL than solid glass at frequencies near the coincidence frequency. This improvement is due to damping (dissipation of vibrational energy) by the plastic inter-layers. However, damping is normally temperature-dependent, and at typical Indian winter temperatures, the increase in TL due to glass lamination may be drastically reduced.

Using double or triple glazing also increases transmission loss at some frequencies, but the improvement depends on the separation of the layers. As shown in Figures 4.3 and 4.4, the mass-air-mass resonance causes a significant dip in the low or mid frequency transmission loss. For typical factory-sealed double glazing, this resonance falls within the frequency range of 200 to 400 Hz. By increasing the airspace and using heavier glass, the resonance frequency can be shifted to frequencies below the range of interest.

Some typical STC ratings are given in Figure 4.9 as a function of the airspace between the two layers of glass; for each doubling of the airspace, the STC increases by approximately 3. These STC values are for sealed windows; for weather-stripped operable windows, the STC values would typically be lower by 3 to 5. Separately supported double windows have acoustical performance similar to factory-sealed double glazing units with the same airspace. To reliably achieve STC values above 45, the use of acoustical windows with specially designed frames and glass mounting is recommended.

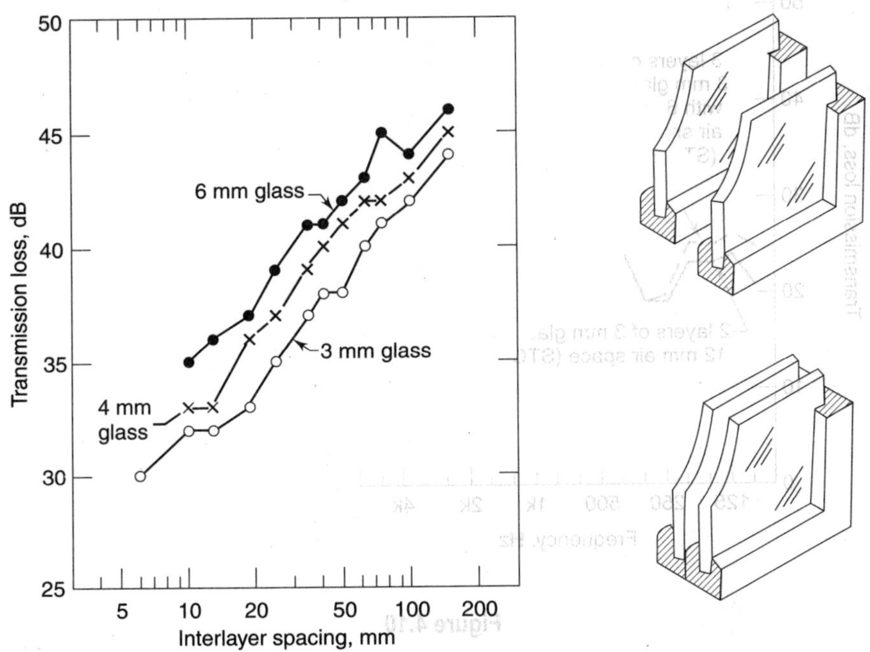

Figure 4.9

Transmission of sound energy through the window frame can significantly reduce the TL, especially for windows with high transmission loss. Windows with lightweight metal frames tend to have lower STC values than those given in Figure 4.9, apparently due to vibration transmitted by the window frame. In general, the use of lightweight frames should be avoided. If very high TL is required, using separate frames supported by structurally independent walls eliminates transmission through the frame. Despite the widespread belief that adding another layer of glass must be beneficial, triple glazing provides essentially the same noise reduction as double glazing, unless the inter-layer separation is very large. Figure 4.10 compares TL data for a double-glazed window with that for a triple-glazed window with the same total airspace. Below the mass-air-mass resonance frequency (about 250 Hz in this case) the TL of triple glazing is about 3 dB higher, consistent with the mass law prediction for increasing the mass by

50%. At higher frequencies, the TL curves are almost identical. The STC ratings for the two windows are the same. The small difference between the TL of double and triple glazing is not restricted to windows whose total thickness is small. Unless both airspaces are significantly larger than 25 mm, the STC of triple windows is very similar to that of double windows.

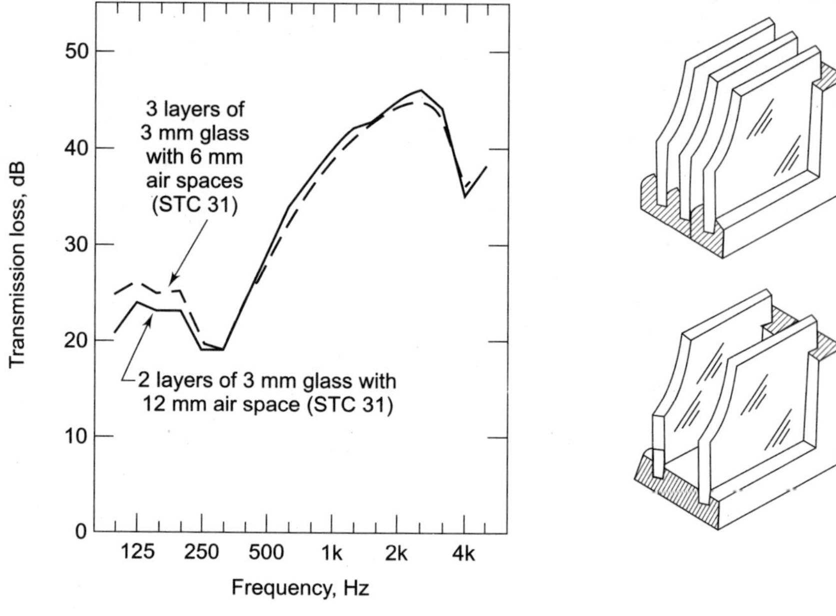

Figure 4.10

4.4 DOORS

The sound insulation provided by a door depends not only on the type of door but also on correct installation. Table 4.3 presents STC values for some common types of interior doors. The first column of STC data illustrates the noise reduction in doors with typical gaps between the edges and the frame (approximately 6 mm total for top and bottom and 3 mm total for the sides). With larger gaps, the STC would be lower. Unless leakage around the door is reduced, the effective STC will not greatly exceed 20, regardless of improvements to the door panels.

The solid curve in Figure 4.11 shows the transmission loss for a typical unsealed solid-core wood door. Even replacing this door panel with solid lead would not increase the STC substantially-the main problem is transmission of sound through cracks around the perimeter of the door.

Table 4.3 Sound Insulation of Conventional Doors

Door Type	Surface Mass kg/m²	STC (Not Sealed)	STC (Good Seals)
Hollow-core wood	7	17	20
Solid-core wood	20	20	28
Hollow-core steel (18 ga.)	25	20	30
Communicating doors (2 hollow-core wood doors, 100 mm space)	7 each	22	26
Communicating doors (2 solid-core wood or hollow steel doors, 70 mm space)	20 each	28	40
Communicating doors (2 solid wood or hollow steel doors, 70 mm space with absorption)	20 each	40	44
Communicating doors (2 solid wood or hollow steel doors, 230 mm space with absorption)	20 each	42	50

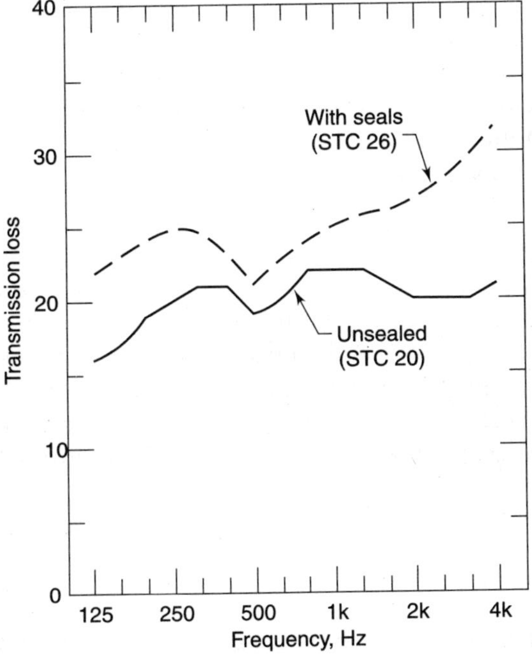

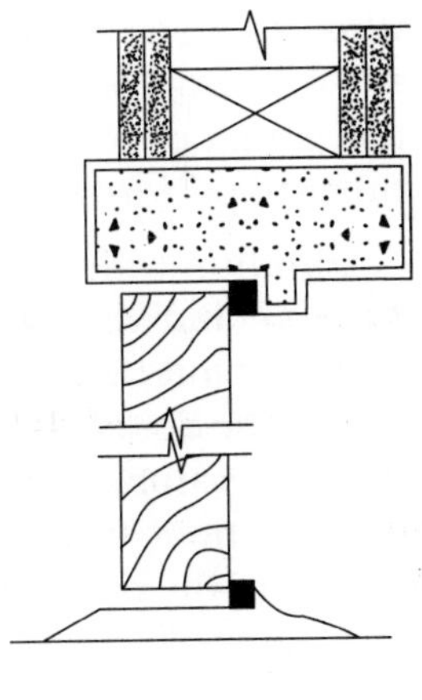

Figure 4.11

The dashed curve in Figure 4.11 shows the increase in transmission loss when gaskets are added around the perimeter of the door. The dip near 500 Hz is the coincidence dip due to stiffness of the solid wood panel. Transmission loss data for most doors resemble the dashed curve here. Conventional steel doors also behave like a single layer, although their coincidence dip is usually not so sharp. STC values in the range from 25 to 30 are typical for conventional steel or solid wood doors with properly installed seals.

Typical STC values for doors with good acoustical seals are presented in the last column of Table 4.3. Rubber or neoprene gaskets compressed between the door and frame are quite effective, but may increase the effort needed to close the door. Magnetic seals (like those used for refrigerator doors) work well with metal doors. If a flat door sill is essential, a 'drop seal' door bottom may be used. To be effective, the seal must press firmly against the sill when the door closes, and must mate well with the bottom edge of the door and the gaskets at the sides of the door. Unless carefully adjusted, drop seals do not provide as good a seal as a compressed gasket or a magnetic seal.

If the acoustical seals are adequate, single doors can achieve STC ratings of 35 or higher. Many manufacturers produce 'acoustical doors' with STC values from 35 to over 50. The sound insulation for specific products can be found in manufacturers' literature. Installation is of critical importance in achieving such high insulation; the rated performance will not be achieved unless the door is properly sealed and the door frame is fitted into the wall so as to avoid cracks or hollows that could transmit sound. Periodic replacement or readjustment of the seals may be required to maintain the acoustical performance of such doors.

Communicating doors (two doors with airspace between) are a less expensive and acoustically more effective alternative in situations where they do not interfere unacceptably with access. Adding acoustical absorption between the doors (a 25 mm layer on one or both of the door surfaces facing the cavity) can provide STC values of 45 or higher (as shown in Table 4.3). With absorptive treatment of the cavity, communicating doors are much less sensitive to the perfection of the perimeter seals than are single 'acoustical' doors.

4.5 FLOORS

There are probably more variations in the acoustical design of floors than there are for walls, because in the design of floors, both impact and airborne noise must be controlled.

Figure 4.12 shows a typical joist floor. From an acoustical point of view, the material used for the joists, whether solid or truss constructions of wood, steel or even concrete, is not important. Most of the weight of a floor system must rest on top of the joists. To obtain an STC appreciably above 50, the upper layer of the floor should have a mass per unit area of at least 50 kg/ m^2. This would require at least 25 mm of concrete and considerably more of a less dense material, such as plywood.

Basic floor components

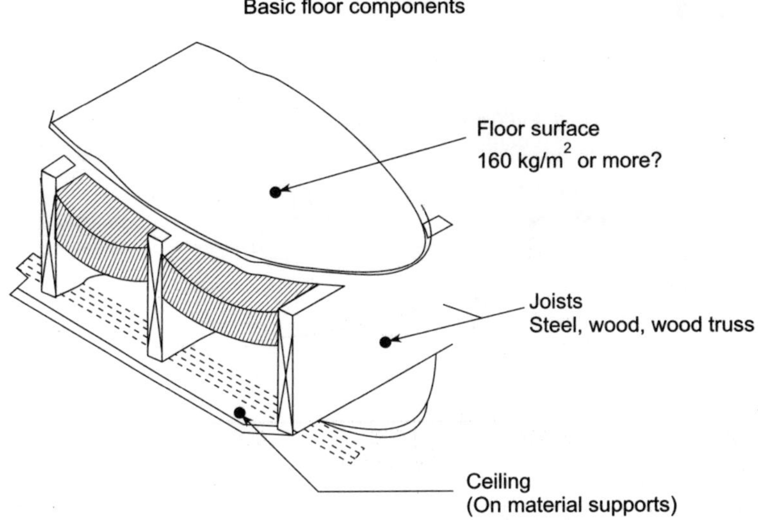

Figure 4.12

For many floors, the noise reduction is limited by sound energy transmitted through the structural connections between the surfaces of the floor and the ceiling below. Providing independent joists to support the ceiling could eliminate this problem, but this solution is generally rejected as being too expensive. A common solution is to use resilient channels to attach the gypsum board to the bottom of the floor joists. The cavity between the underside of the floor and the ceiling is normally large enough so that mass-air-mass resonance may be ignored. If rigid structural connections between the surfaces are avoided, adding absorptive material to the spaces between the floor joists can give further improvements in transmission loss, similar to those listed for walls in Table 4.2. (Fastening one layer of gypsum board directly to the bottom of the floor joists and then using resilient channels to support a second layer is not a good way of reducing the structural connection. The mass-air-mass resonance caused by the small airspace between the layers of gypsum board substantially reduces the low frequency transmission loss.)

In many buildings, the benefit of cavity absorption and resilient attachment of the ceiling is minimal because of sound energy transmitted through the supporting structure. Resiliently suspending the ceiling below the floor does not reduce either transmission along the floor to adjacent rooms on the same level or the transmission down the supporting walls. The most effective means of reducing the transfer of sound energy into the structure is the use of a 'floating floor'. A layer of concrete supported on a resilient layer or special resilient mounts is one common form of floating floor. Some approaches to resilient support of floors are illustrated in Figure 4.13.

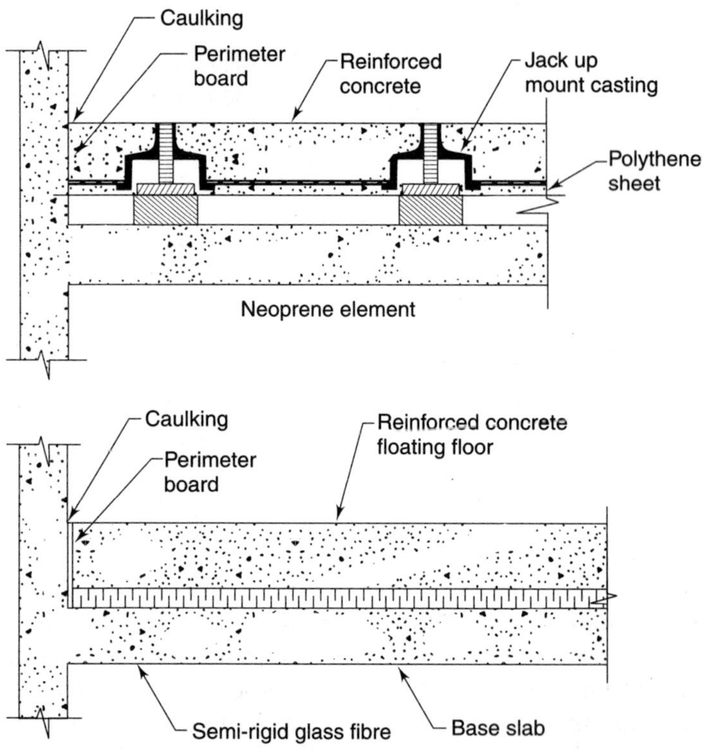

Figure 4.13

The design of floating floors is not trivial. The top surface must be heavy and rigid enough to keep motion acceptably low when it is walked on, and the resilient support must have appropriate stiffness. These details require specific design advice from an expert. Although floating floors are not inexpensive, they do permit design of floor systems with very high transmission loss for airborne noise and good resistance to impact noise, as discussed below.

Control of Impact Noise

Impact sound, in particular, footstep noise, is one of the most difficult forms of acoustical energy to control. When an airborne sound wave strikes a wall, only a small part of its sound energy is transferred to the wall. The transfer of energy when a solid object strikes a floor or a wall is a much more efficient process.

The physical parameters that control the transmission of impact sound through a floor system are basically the same as those that control airborne sound. Lightweight floors are easily set into vibration and resonate at low frequencies; when walked on they emit annoying booming sounds. Increasing the mass of the floor layers reduces the response to the force and therefore reduces the transmitted sound.

Dividing a floor into two isolated or resiliently connected layers decreases the transmission of vibration between them and thus reduces both the impact and airborne sound radiation on the other side. Sound-absorbing material inside properly isolated floor cavities has the same beneficial effect for impact noise as shown earlier for airborne sound reduction.

An additional way of controlling impact noise, which is not effective for airborne is to lay a soft resilient layer, such as a carpet, on top of the floor. The impact force is cushioned by the resilient layer, reducing the vibrational energy that is transferred to the floor structure. For example, the addition of carpet and underlay to a 150 mm slab of concrete increases the impact insulation class (IIC) from about 40 to 65 or more, depending on the type of carpet. In contrast, adding a carpet and underlay has little effect on the STC of a floor. Impact noise can also be effectively controlled by the use of a floating floor. If the top surface must be hard (as in a kitchen or bathroom) then a floating floor is the most effective way of dealing with impact sound.

The IIC is generally used to categorize the effectiveness of floors in preventing transmission of impact noise. Much of the sound emitted when lightweight floors are walked on is at frequencies below the range used for calculating the IIC. Also, the sounds generated by the standard hammer machine used to measure the IIC are quite different from sounds made by human walkers on this type of floor. With some lightweight floors it is possible to get a high IIC rating by adding a carpet, although generation of low frequency noise when the floor is walked on could still cause serious annoyance. Until a better test is invented and adopted in building codes, one can do little except to use the IIC ratings, with the awareness that they may be rather optimistic about the impact performance of lightweight floors.

4.6 SUMMARY

In designing to prevent sound transmission, all leaks must first be eliminated. The next factor usually considered is the mass of the partition; an increase in weight generally increases the transmission loss. However, this improvement is limited by the coincidence dip, which depends on the material's stiffness and thickness.

Cavity constructions are a practical alternative to massive solid constructions. Reducing structural connections by which vibrational energy can be transferred across the cavity is the first priority. Once this is accomplished, increasing the depth of the cavity and adding absorptive material inside it improve the transmission loss. At low frequencies, the mass-air-mass resonance limits the transmission loss. If this resonance is below 80 Hz, it is not a significant problem for airborne noise. The resonance frequency can be lowered by increasing the mass of the surface layers or by increasing their separation.

Impact sound is one of the most difficult forms of acoustical energy to control. In general the most effective way to control impact sound is to use a floating floor or a resilient surface such as carpet to reduce the vibrational energy reaching the structure.

5

ACOUSTICS IN PRACTICE

5.1 INTRODUCTION

When sound transmission through building components is determined in a laboratory, sound travels only directly through the component under test. In buildings, however, other flanking paths around separating assemblies are likely to be involved. This Chapter deals with some of the problems encountered when the components are combined into systems.

To create a successful acoustical design for a building, not only must the performance of the components selected be considered, but also the details of the interconnections, in other words, the system performance. It is difficult to specify quantities in this area, because of the complexity of the problems and the lack of available information. The best way to ensure acoustical success for a building is to select criteria, select components and layout intended to satisfy the criteria, and then verify the acoustical performance achieved by field testing at an early stage in the construction.

5.2 CRITERIA FOR FIELD SOUND TRANSMISSION CLASS AND FIELD IMPACT INSULATION CLASS

A number of interacting factors are important in determining whether an occupant of a multi-family dwelling is bothered by noise from neighbouring units. These are:

1.   the impact and airborne sound transmission characteristics of party walls and floors,

2.   the level of noise generated in neighbouring homes,

3.   the level of background sound in the occupant's own home, and

4.   the sensitivity of the occupants.

For example, if the level of the background sound is high enough, then intruding sounds will be masked and will not be detectable. The last three factors can vary widely and are generally beyond the control of the designer. The pragmatic approach to this situation is to select a value of field impact insulation class (FIIC) or field sound transmission class (FSTC) that will provide protection for most situations and accept those cases where annoyance is caused by unusually noisy neighbours, low background sound levels, or sensitive occupants.

The criteria given here are for complete rooms tested in the field and not for components tested in the laboratory. Laboratory ratings do not apply to systems, only to components. This is emphasized by the use of the letter F in the respective ratings.

For occupants of multi-family dwellings to enjoy a reasonable degree of acoustical privacy, the effective FSTC for walls or floors separating homes should be at least 55 in the finished building. Higher values provide an improved living environment and can be achieved with typical building materials without resorting to extreme designs, although care is needed during the design and construction processes. Minimum values greater than 50 are commonly required in Europe, and in some countries, field measurements ensure that criteria are satisfied.

The field impact insulation class for a hard-surfaced party floor should be at least 55. When a floor is carpeted, the FIIC achieved should be at least 65, a rating that is relatively easy to obtain by including an underlay. Unfortunately, despite having a high FIIC rating, some lightweight floors are unsatisfactory in practice because walking on them generates low frequency noise. The standard test procedures do not evaluate the low frequency behavior of floor systems, for technical reasons. Some experience is necessary with these floors.

Having selected components, the designer has no guarantee that the performance obtained in the finished building will equal that obtained in laboratory tests or in some other building, because the elements of the system can interact in negative ways. To ensure a successful building, acoustical testing should be carried out before occupancy of a test unit in a partially completed building; this can uncover many problems which can then be avoided in subsequent units.

5.3   BUILDING LAYOUT

The acoustical criteria recommended above are for separation of normal spaces such as living or sleeping areas. If a sensitive area such as a bedroom were to be placed next to a noisy machine room, then the FSTC to be achieved would have to be increased to 65 or higher. In this situation, advice from a qualified acoustical consultant should be sought. On the other hand, the wall separating an apartment from the corridor need not be much better than FSTC 50, because the noise source would probably only be voices or attenuated music from an adjoining apartment with a temporarily open door.

Thus the choice of layout for a building has a great deal of influence on the sound insulation criteria that must be selected for the components and the system. One of the simplest means of controlling noise is to separate noise sources from sensitive receiving areas as far as possible. Machine rooms, garbage chutes, elevator shafts and other noisy areas in buildings should be located as far as possible from sensitive living or sleeping areas. Placing relatively quiet areas such as bedrooms next to each other also helps to minimize the noise reduction required for occupant satisfaction.

In Figure 5.1, the bedroom in the apartment on the left is likely to be noisy because of the adjacent elevator shaft and kitchen. In the apartment on the right, the bedroom is remote from the elevator shaft and kitchen, so the only major sources of noise will be the apartments above and below. If this layout is used for apartments side by side, then the floor plan should be reversed in adjacent units to put bedroom next to bedroom and living room next to living room.

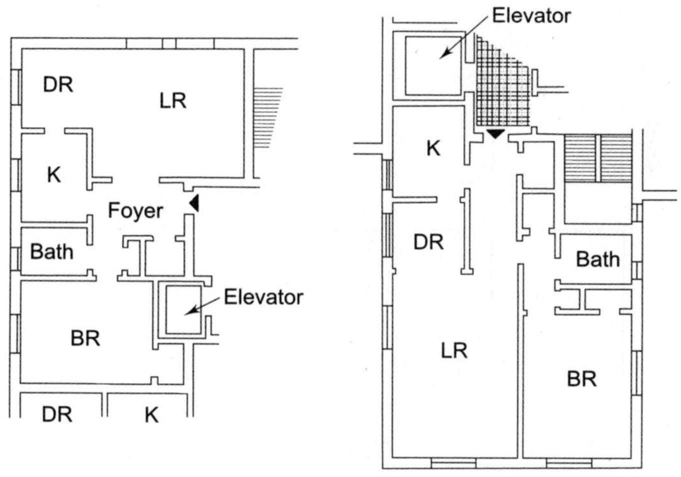

Figure 5.1

Public corridors serve as buffer zones between apartments but the doors into them often permit noise to intrude. This is especially true if the doors are poorly sealed, with large gaps at the bottom.

In Figure 5.2 not only is there a likelihood of noise intrusion into apartments from the corridor but also between adjacent apartments by way of the doors. To reduce this intrusion, doors into the corridor should be sealed with weather stripping so that no significant leaks exist to increase sound transmission.

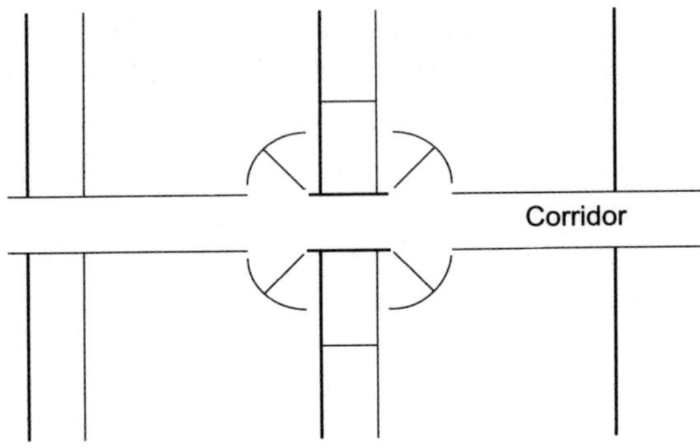

Figure 5.2

A vestibule between the living areas and the corridor (as shown in Figure 5.3) also helps to reduce noise intrusion. This is commonly done in hotels in Europe and greatly reduces the intrusive noise from corridors. This technique might be put to good use in a restaurant to keep kitchen noise out of the dining area.

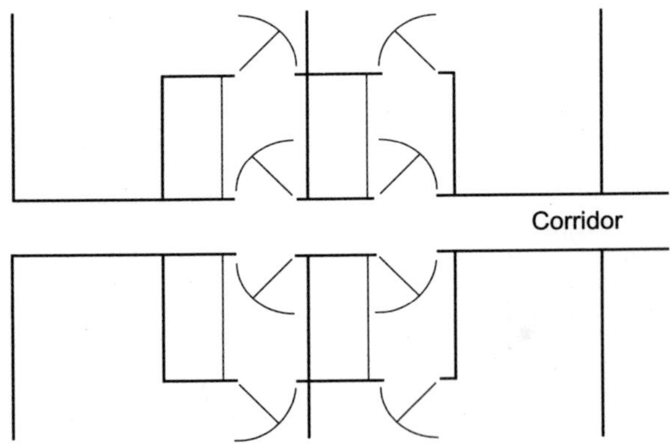

Figure 5.3

Another way to deal with this problem is to offset doors along the corridor (as shown in Figure 5.4) and to add sound-absorbing materials to the surfaces in the corridor to reduce noise transmission along it. Materials may conveniently be added to the ceiling or upper parts of the walls of the corridor or a carpet can be put on the floor.

Once the preliminary building layout has been selected, the acoustical requirements for the walls and floors separating the different spaces should be examined. Where the layout requires the use of walls or floors with very high sound transmission loss ratings, the rooms in the building can be re-arranged so that the acoustical requirements become less restrictive.

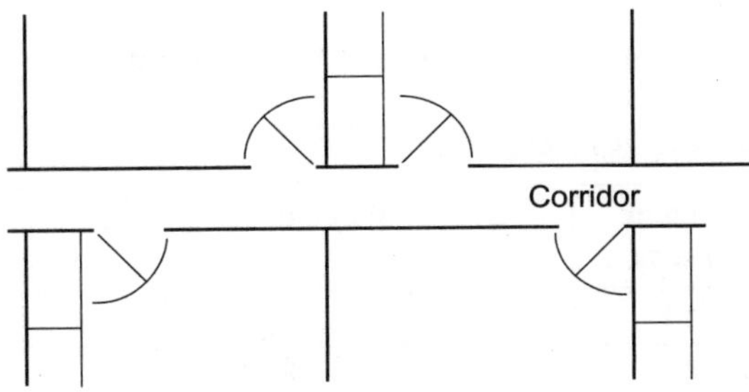

Figure 5.4

5.4   MULTI-ELEMENT TRANSMISSION

It is often necessary to combine walls with other elements such as doors or windows. The latter usually do not have high sound transmission losses and reduce the overall sound insulation of the system. Thus the problem facing the designer is to combine two or more components with different areas and different sound transmission losses so that the overall sound transmission losses and STC rating for the system meet the design criterion.

The sound energy passing through each component depends on its area and sound transmission loss. Transmission loss is a function of frequency, so calculations should be carried out for each frequency band of interest and then the STC calculated for the composite transmission loss spectrum. However, for a first look at a situation, it is sufficient to work only with the STC ratings.

A worksheet for doing these calculations is given at the end of this chapter. A simplistic approach to this problem of combining components is to require that all components have the same rating. In practice this is usually not possible when the required STC is fairly high and one has to deal with windows and doors. A more practical and more economical approach is to keep the area of the weaker components small and to allow them to pass more of the sound energy. This is compensated for by reducing the sound energy passing through the stronger component, usually the wall.

Table 5.1 gives an example of the use of the worksheet for a wall, door and window with areas of 82%, 15% and 3% of the total. A good starting point for the partitioning of the energy among the components is to assume that each carries the same fraction of the total (in this case, 33% for each). This sometimes does not produce a satisfactory design, and in Table 5.1 the percentage of energy transmitted through each component is set to 10%, 45% and 45%, for the wall, door and window, respectively. The weaker components are allowed to carry more of the energy. The required combined STC is 50 and the calculations show that the window should have STC 38, the door STC 45 and the wall, STC 59.

Table 5.1 Calculation of Component STC for a Wall, Window and Door

Required combined STC: 50 = STC′					
Component	Percent energy	C1	Percent area	C2	Component STC = STC′ − C1 + C2
1. Wall	10	−10	82	−1	59
2. Door	45	−3	15	−8	45
3. Window	45	−3	3	−15	38

The same example, with the window eliminated from the design and the door passing 90% of the sound energy, is presented in Table 5.2. The wall still needs to have an STC of 59 and the door an STC of 42.

Table 5.2 Calculation of Component STC for a Wall and Door

Required combined STC: 50 = STC′					
Component	Percent energy	C1	Percent area	C2	Component STC = STC′ − C1 + C2
1. Wall	10	−10	85	−1	59
2. Door	90	0	15	−8	42

Reducing the energy going through the wall to only 1% would produce no benefit: the door STC would still have to be 42, even though the wall would need an STC of 69. This apparent discrepancy is due to the logarithmic nature of the calculations.

The important point to realize from these calculations is that there is a limit to what can be done by strengthening one component relative to the others. All too soon the overall STC is determined by the sound transmission through the weak component. The calculations also show the importance of looking at the requirements for the system before construction. It is possible to construct a window that has an STC rating of 38 from readily available components but single doors with STC's greater than 40 would be classified as 'acoustical' doors and would be quite expensive. A pair of doors in the same frame could be constructed to give a high enough STC but double door system might not be acceptable. It might be more economical to rework the design so that the acoustical requirements on this system are less demanding, for example, by introducing a vestibule as a buffer.

It is sometimes necessary to start with readily available components of known ratings and areas and to calculate the combined rating to ensure that it is enough for some particular use. Part II of the worksheet can be used for this calculation. For example, an STC 50 wall with an area of 10 m^2 combined with an STC 25 door with an area of 2 m^2 would give a system with an STC of 33 (as shown in Table 5.3).

Table 5.3 Calculation of Combined Sound Transmission Class for Wall and Door with Known Values

Working assumed combined STC: 35 (STC')					
Component	STC	Percent area	C2	CI = STC' − STC + C2	Percent energy
1. wall	50	84	−1	−16	2.5
2. window	25	16	−8	2	160

Percent sum of energies = 162.5

Actual STC = STC - C' = 35 − 2 = 33
(See the Worksheet, Part II) C' = 2 dB

5.5   FLANKING PATHS

When partitions or floors are tested in acoustical laboratories, great care is taken to ensure that the only significant sound transmission path is through the test specimen itself which, ideally, is completely isolated from the rest of the structure. Solid connections between the specimen and the surrounding structure are eliminated where possible or made sufficiently resilient.

To approach the same sound attenuation in buildings that is attained in laboratory tests, a serious attempt has to be made to ensure that energy transmission along flanking paths is reduced to a minimum by introducing breaks in the construction and resilient connections. In the ideal situation, each apartment would be an independent, resiliently suspended unit within the building, like a set of boxes within a larger box.

In a building where all the components are rigidly connected, sound energy can be transmitted through the ceilings, walls and floors of the structure to adjacent rooms; vibrations in the structure cause sound to be radiated from all surfaces as shown in Figure 5. Those paths that are not direct are flanking paths.

Once sound has entered a structure and is propagating as vibration, it can travel for considerable distances with only slight energy losses. The energy loss depends on the material and on the construction details. The losses at junctions where floors and walls meet can be predicted for simple monolithic structures like solid concrete, but only qualitative estimates can be made for more complicated junctions. The vibrations in the material are continually re-radiating energy as sound from both sides of the surface. These paths have much more serious consequences for impact sound transmission than for airborne sound transmission because the transfer of impact energy to the structure can be very efficient when surfaces or impactors are hard.

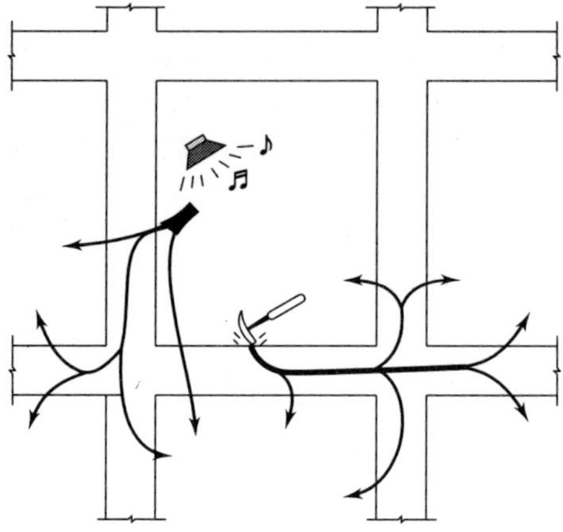

Figure 5.5

If for structural reasons a floor is rigidly connected to the adjoining and supporting walls, impact sound transmission along flanking paths can be a serious problem. Resiliently suspending the ceiling below the floor, as in Figure 5.6, does not reduce in any way the transmission paths to adjacent rooms on the same level down the walls.

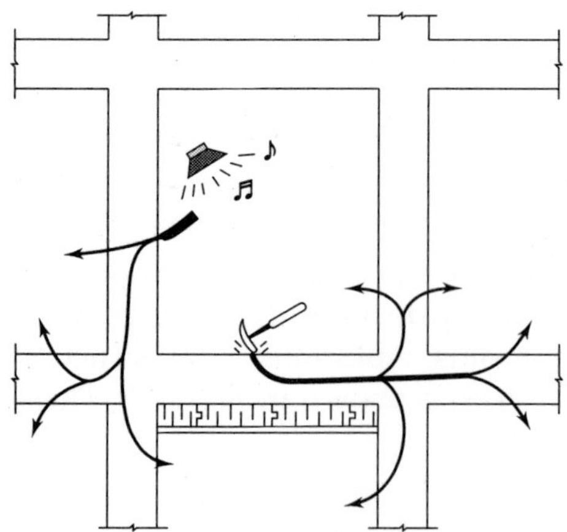

Figure 5.6

A much better way to reduce impact sound transmission along flanking paths is to use a floating floor, shown in idealized form in Figure 5.7. Sound energy will be dissipated within the floating slab or substantially reduced as it propagates from the floor slab through the resilient layer or supports to the rest of the structure.

Figure 5.7

5.6 FLANKING SOUND TRANSMISSION IN OFFICE PLENUMS

A common flanking sound transmission problem in of office buildings arises from the use of a continuous suspended ceiling and partitions that do not extend from floor slab to floor slab. This creates an easy path for sound between offices via the ceiling and the plenum space (Figure 5.8). Typically, leaks in the ceiling and around the partition allow easy passage for the sound; the sound insulation between the offices is usually determined by these leaks and the flanking paths, rather than by the partitions.

The attenuation of sound propagating through the plenum must be increased by alterations to the ceiling or to the space above the partition. The gap between the top of the wall and the ceiling slab may be blocked with some easily worked material such as wood fiberboard, gypsum board, plywood or perhaps mass-loaded vinyl. The partition may simply be extended to the slab above but this is often impractical. Good results have been obtained when the space above the wall was filled with bats of glass fibre to form a solid block about 1.2 m thick. It is always important to ensure that there are no residual leaks; the junction where the ceiling meets the wall is particularly critical. Continuous cavities such as heating units, light back-to-back power outlets and short runs of straight unlined duct work should be avoided. A potential difficulty in such of offices is that changes made to improve acoustical privacy may interfere with air distribution. This possibility should not be overlooked.

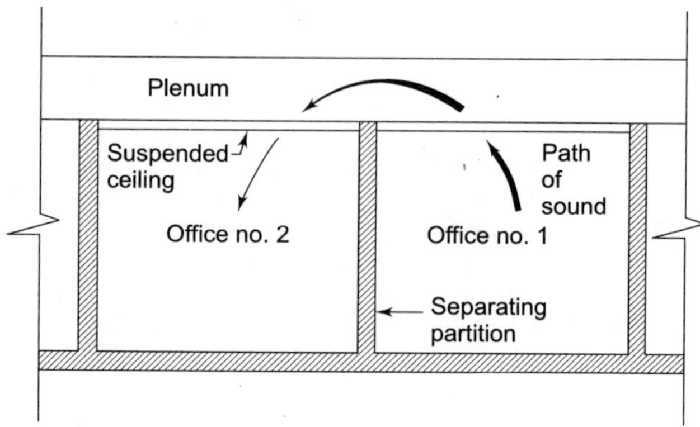

Figure 5.8

5.7 CONSTRUCTION DETAILS

The details of the construction where walls and floors join can greatly influence the sound transmission along flanking paths. In the situation shown in Figure 5.9, both the wall and the floor tested in a laboratory attained high STC ratings. The sound isolation between horizontally adjacent homes is seriously reduced, however, because of the transmission of vibration along the top surface of the floor. The wall is 'short-circuited' by the top layer of the floor. Impact noise transmission along the floor into the adjacent apartment is a particularly bad problem. The remedy for this is to introduce a structural discontinuity (a vibration break) to reduce the passage of vibration along the lightweight layer without compromising the structural integrity of the system. Such a break is shown on the right in Figure 5.10, where a similar problem can exist in a floating floor.

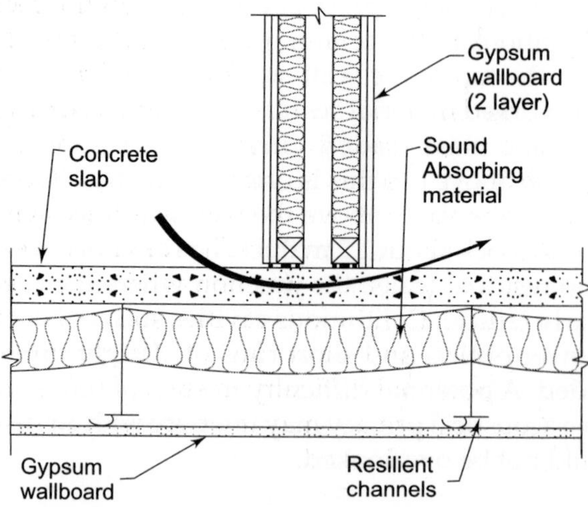

Figure 5.9

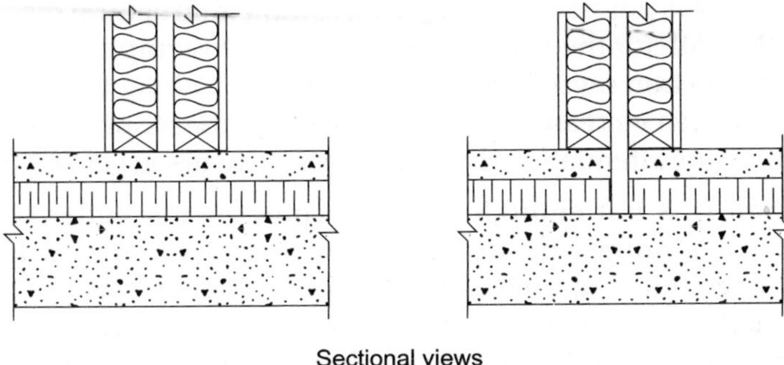

Sectional views

Figure 5.10

Similar flanking paths arise in wall systems; the principles remain the same—some kind of vibration break should be introduced.

Figure 5.11(a) shows a horizontal section through a wall junction where there is a risk of flanking transmission along the wallboard. Figure 5.11(b), where the wallboard is not continuous, is preferable for acoustical purposes. One reason for using the construction in 5.11(a) might be to ensure a more airtight construction on an outside wall. If this is the case, the detail in 5.11(c) can be used to ensure that both acoustical and air tightness requirements are satisfied. A small gap between the layers of wallboard is caulked and covered with a flexible tape. This resilient connection is enough to reduce the transmission of vibration along the surface.

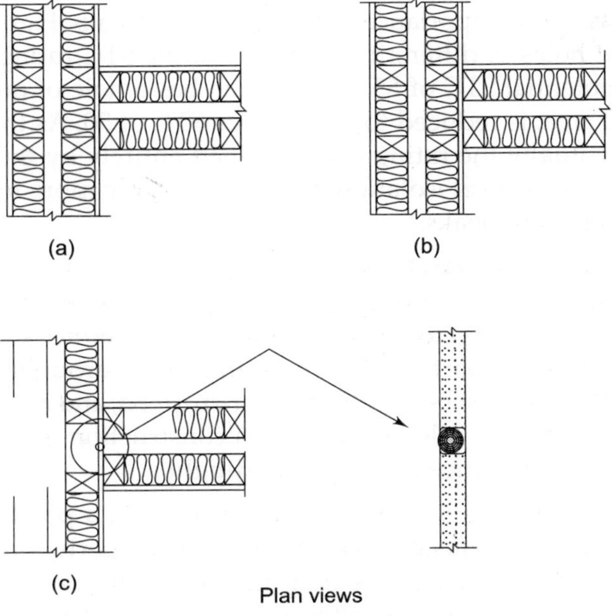

(a)

(b)

(c)

Plan views

Figure 5.11

5.8 PLUMBING NOISE

Plumbing noise can be regarded as a special case of machinery noise. Three factors affect the noise in plumbing systems: water pressure, water velocity, and the number and types of constrictions and fittings. Turbulence in the water is a source of hydrodynamic noise. This can be reduced by selecting well designed fixtures and minimizing fittings, sharp bends and constrictions in the system. The static water pressure in the system should be kept below about 0.25 MPa, using a pressure regulating valve if necessary. Water flow rates should be about 2 m/s or less. Increasing the diameter of the piping helps to reduce the flow rate and the noise, but this is less effective than reducing the pressure in the system.

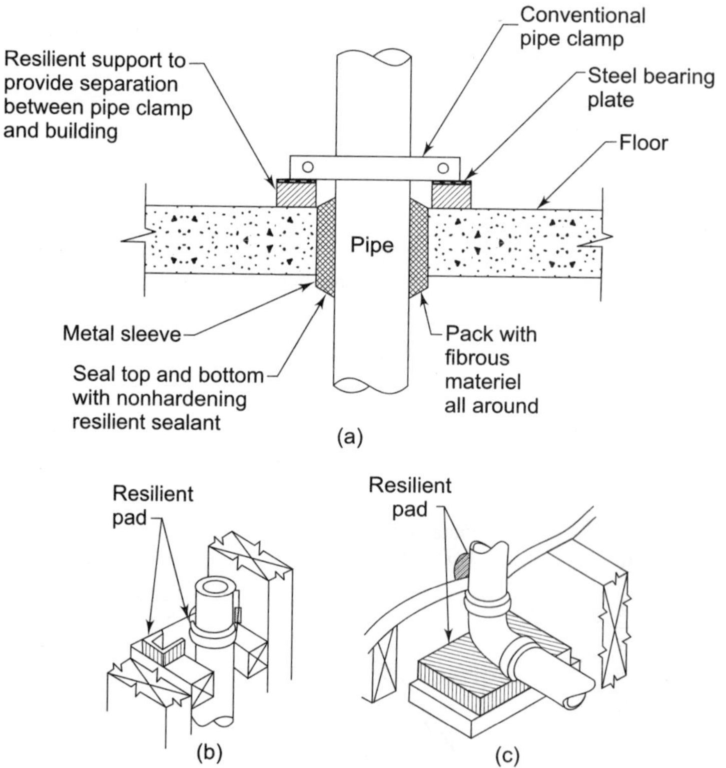

Figure 5.14

Plumbing noise is almost always transmitted as structure-borne vibration and eventually radiates from lightweight surfaces in many different places. The most common error made is to connect the water supply and waste pipes and other fixtures rigidly to the structure. To reduce the amount of noise entering the structure, pipes should be isolated from it, using resilient sleeves and hangers, and flexible connections. Examples of how to do this are given in Figures 5.14 (a), (b) and (c). Where pipes penetrate sound barriers, the holes should be carefully sealed as shown. Contact with lightweight surfaces, which radiate sound easily, must be avoided at all costs.

Water hammer occurs when valves are closed quickly, producing a shock wave in the system which causes the pipes to vibrate. Reducing pressure and velocity, and avoiding quick-closing valves, help to reduce water hammer. Air-filled stubs can be introduced in the piping close to appliances such as washing machines; they act as springs to reduce the shock (as shown in Figure 5.15).

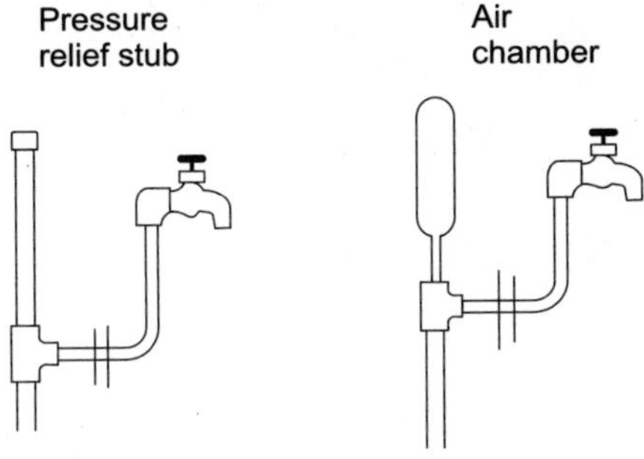

Figure 5.15

Putting all of these ideas into practice requires care and attention to detail. Two complete layouts are shown in Figures 5.16 and 5.17, for a bathtub and a toilet.

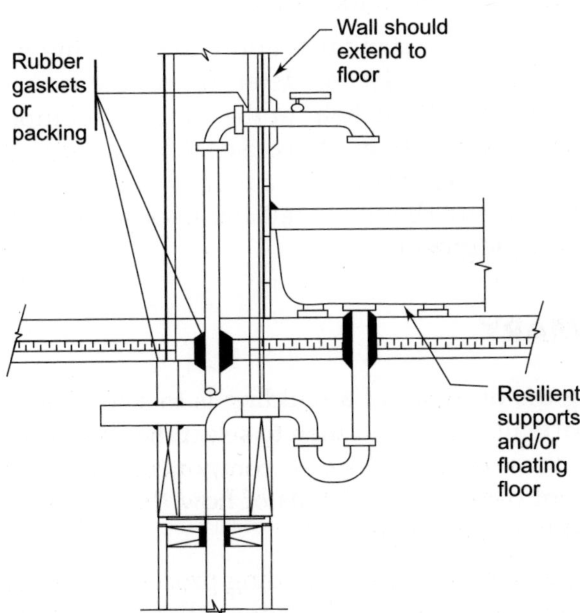

Figure 5.16

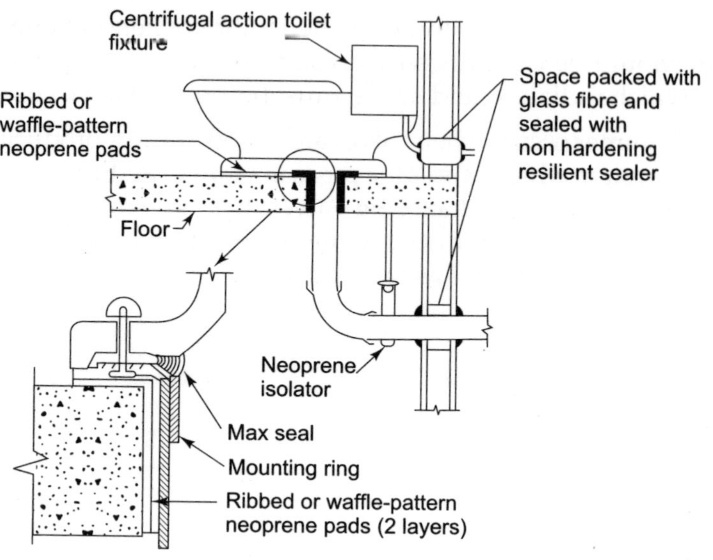

Figure 5.17

As with other noise sources, the layout in the building is important for control of plumbing noise. Where possible, pipes should not be installed in common walls or in floor-ceiling systems adjoining areas where quiet is important tithe residents. In critical situations pipes should be enclosed in pipe chases heavy enough to provide good noise reduction and with sound-absorbing materials on the inside.

In India, no acoustical standards apply to plumbing noise at present. In Europe, test methods exist for determining the noise generated by typical components and these help in the choice of components for new construction. The present state of acoustical knowledge does not allow the prediction of the level of sound in a room when a particular plumbing system is combined with different wall and floor systems. One has to rely on experience and judgment.

5.9 SUMMARY

The task facing the designer of a building is to select the criteria that are relevant in a particular case and then to select the components and layout that will satisfy those criteria. Once the components have been selected, the design task is not over; all the details of how the components will interconnect must still be considered.

On-site inspection during the building process will often reveal errors that can be easily corrected at the time. All too often these errors are concealed and cannot be easily uncovered or repaired in a finished building.

Acoustical testing in a partially completed building ensures that errors are not needlessly repeated and gives the opportunity for correction. In all of this process, professional help makes the designer's life much easier and helps produce a building that will be quieter and more satisfactory to its occupants.

Worksheet

Part 5.1 For Calculating Component STC from a Required Design STC and the Component Areas

Design STC:______ = STC′					
Component	Percent energy	C1	Percent area	C2	Component STC = STC′ −C1 + C2
1._____	_____	_____	_____	_____	_____
2._____	_____	_____	_____	_____	_____
3._____	_____	_____	_____	_____	_____
4._____	_____	_____	_____	_____	_____

Table for converting percentages to decibel factors; C = 10 log (Percent/100) [Use centre column when converting from C to %, and select range of % when converting from % to C]

Percent	C	Percent	C	Percent	C
560 (630) 700	8	56 (63) 70	−2	5.6 (6.3) 7.0	−12
446 (500) 560	7	45(50)56	−3	4.5 (5.0) 5.6	−13
357 (400) 445	6	36 (40) 44	−4	3.6 (4.0) 4.5	−14
281 (315) 356	5	28 (32) 35	−5	2.8 (3.2) 3.6	−15
223 (250) 280	4	22 (25) 27	−6	2.2 (2.5) 2.8	−16
178 (200) 222	3	18 (20) 22	−7	1.8 (2.0) 2.2	−17
143 (160) 177	2	14 (16) 17	−8	1.4 (1.6) 1.8	−18
111 (125) 142	1	11 (12) 14	−9	1.1 (1.2) 1.4	−19
89 (100) 110	0	9 (10) 11	−10	0.9 (1.0) 1.1	−20
71 (80) 88	−1	7 (8) 9	−11	0.7 (0.8) 0.9	−21

Instructions for Use

1. Divide the energy among the components. A good starting point is to assume equal energy for each one.

2. Select the values of C from the conversion table (use the centre column) and place them in the column headed C1.

3. Enter the component areas in the next column.

4. Select the values of C from the conversion table (use the centre column) and place them in the column headedC2.

5. Calculate the sum STC ' - C1 + C2 for each row in the table taking account of the signs. These values are the required STCs for each component.

The calculation can be repeated with larger fractions of the energy going through the weaker components if the design is not satisfactory.

Worksheet

Part 5.2 For Calculating total STC from Component Area and STC

Working assumed STC:_____ (STC')					
Component	STC	Percent area	C2	C1 = STC' − STC + C2	Percent energy
1._____	_____	_____	_____	_____	_____
2._____	_____	_____	_____	_____	_____
3._____	_____	_____	_____	_____	_____
4._____	_____	_____	_____	_____	_____
			Percent sum of energies = ___ C'→ = ___ Actual STC total = STC' − C' =		

Instructions for Use

1. Assume a value for the combined STC (the arithmetic average is convenient).

2. Enter the known values of STC and percent area for each component in the table rows.

3. Select the value of C appropriate to the areas from the conversion table. (Select the range in which the area lies.)

4. Calculate the values of C1 for each row.

5. From the conversion table, select the value of percent energy corresponding to the value of C1 in the conversion table (use centre percent column).

6. Add the percent energies and read the value of C ' from the conversion table.

7. The actual STC for the system is given by STC' – C ' taking account of signs.

6

ROOM ACOUSTICS

6.1 INTRODUCTION

Much has been written in the popular and professional audio press about the acoustic treatment of rooms. The purpose of such treatment is to allow us to hear more of the loudspeaker and less of the room. I am convinced that a properly designed sound system can perform well in a great variety of rooms and requires only a minimum of room treatment if any at all.

To understand this claim let's look at the typical acoustic behavior of domestic size listening rooms, which have linear dimensions that are small compared to the 17 m wavelength of a 20 Hz bass tone, but are acoustically large when compared to a 200 Hz or 1.7 m wavelength midrange tone (G_1 on the piano keyboard).

Below 200 Hz the acoustics of different locations in the room are dominated by discrete resonances. Above 200 Hz these resonances become so tightly packed in frequency and space that the room behaves quite uniformly and is best described by its reverberation time RT_{60}.

Room treatment can be very effective above 200 Hz, but the same result may be obtained more aesthetically with ordinary furnishings, wall decoration, rugs on the floor and the variety of stuff we like to surround ourselves with. How much treatment is needed, or how short the reverberation time should be, depends on the polar radiation characteristics of the loudspeaker. For my open baffle speaker designs a room becomes too dead when its RT_{60} falls below 500 ms.

We can think of sound as propagating like a light ray. Thus, we can use a mirror to find the region on the side wall or ceiling where sound from response. Practical loudspeakers are neither pure monopoles nor pure dipoles except at low frequencies where the acoustic wavelengths are large compared to the cabinet dimensions.

The ideal monopole is omni-directional at all frequencies. Very few speaker designs on the consumer market approach this behavior. This type of speaker illuminates the listening room uniformly and the perceived sound is strongly influenced by the room's acoustic signature. The result can be quite pleasing, though, because a great deal of acoustic averaging of the sound radiated into every direction takes place. The speakers tend to disappear completely in the wide sound field. Unfortunately, the direct sound is maximally masked by the room sound and precise imaging is lost, unless the listening position is close to the speakers.

The typical box speaker, whether vented, band-passed or closed, is omni-directional at low frequencies and becomes increasingly forward-directional towards higher frequencies. Even when flat on-axis, the total acoustic power radiated into the room drops typically 10 dB ($10x$) or more between low and high frequencies. The uneven power response and the associated strong excitation of low frequency room modes contributes to the familiar and often desired generic box loudspeaker sound. This cannot be the avenue to sound reproduction that is true to the original.

The directional response of the ideal dipole is obtained with open baffle speakers at low frequencies. Note, that to obtain the same on-axis sound pressure level as from a monopole, a dipole needs to radiate only 1/3rd of the monopole's power into the room. This means 4.8 dB less contribution of the room's acoustic signature to the perceived sound. It might also mean 4.8 dB less sound for your neighbor, or that much more sound to you. Despite this advantage dipole speakers are often not acceptable, because they tend to be constructed as physically large panels that interfere with room aesthetics, and they seem to suffer from insufficient bass output, critical room placement and a narrow "sweet spot".

These claims are true to varying degree depending on the specific design of a given panel loudspeaker. Because of the progressive acoustic short circuit between front and rear as the reproduced signal frequency decreases, the membrane of an open-baffle speaker has to move more air locally than the driver cone of a box speaker for the same SPL at the listening position. This demands a large radiating surface area, because achievable excursions are usually small for electrostatic or magnetic panel drive. The obtained volume displacement limits the maximum bass output. Non-linear distortion, though, is often much lower than for dynamic drivers. Large radiating area means that the panel becomes multi-directional with increasing frequency which contributes to critical room placement and listening position.

If the open-baffle speaker is built with conventional cone type dynamic drivers of large excursion capability, then adequate bass output and uniform off-axis radiation are readily obtainable in a package that is more acceptable than a large panel, though not as small as a box speaker. This type of speaker has a much more uniform power response than the typical box speaker. Not only is its bass output in proportion to the music, because room resonance contribution is greatly reduced, but also the character of the bass now sounds more like that from real musical instruments. My hypothesis is that three effects combine to produce the greater bass clarity:

1.   An open baffle, dipole speaker has a figure-of-eight radiation pattern and therefore excites fewer room modes.

2.   Its total radiated power is 4.8 dB less than that of a monopole for the same on-axis SPL. Thus the strength of the excited modes is less.

3.   A 4.8 dB difference in SPL at low frequencies is quite significant, due to the bunching of the equal loudness contours at low frequencies, and corresponds to a 10 dB difference in loudness at 1 kHz.

Thus, bass reproduced by a dipole would be less masked by the room, since a dipole excites fewer modes, and to a lesser degree, and since the perceived difference between direct sound and room contribution is magnified by a psychoacoustic effect.,

The off-axis radiation behavior of a speaker determines the degree to which speaker placement and room acoustics degrade the accuracy of the perceived sound. Worst in this respect is the typical box speaker, followed by the large panel area dipole and the truly omni-directional designs. Least affected is the sound of the open-baffle speaker with piston drivers.

Often concern is expressed over the fact that the rear radiation from a dipole is out of phase with the front radiation, and that thus any sound reflected from a wall behind the speaker would cancel sound coming from the front of the speaker. Cancellation can only occur when direct and reflected sounds are exactly of opposite phase (180 degrees) and of the same strength. Since direct and reflected sounds travel paths of different length, they undergo different amounts of phase shift. Thus, the phase and magnitude conditions for cancellation are given only at certain frequencies, if at all. At some other frequencies direct and reflected sounds will add. The same also applies to a monopole speaker in front of a wall. The only difference is in the frequencies for which addition and subtraction occur. The best remedy is to move the speaker away from the wall, or to make the wall as sound absorptive or diffusive as possible.

6.3   ROOM REVERBERATION TIME T_{60}

Reverberation time is the single most important parameter describing a room's acoustic behavior. The following discussion might get a little technical but will illustrate how sound builds up and decays in a room and the effect it has upon clarity of reproduction.

Sound Waves between Two Walls

Take the example of a speaker in a wall and a second wall at distance L in front of it. As the cone vibrates it will send out an acoustic wave which gets reflected back by the second wall, returns to the first wall, gets reflected again back to the second wall and so on. If the frequency of vibration is such that the distance L corresponds to half of a wavelength, then the cone movement is in phase with the reflected wave and the sound pressure keeps building up. Eventually equilibrium is reached between the energy supplied by the cone movement and the energy absorbed by the two walls and the air in between.

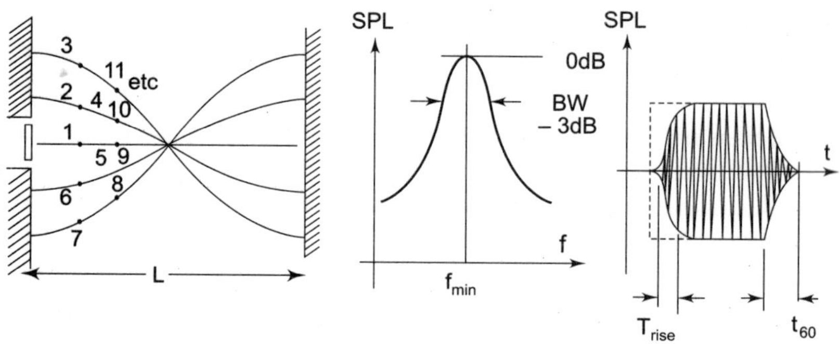

Figure 6.2

This is a standing wave resonance or mode condition and if we change the frequency of cone vibration, we trace out the resonance curve that is typical for any simple system containing mass, compliance and energy loss. As frequency is increased another resonance occurs when L equals to a full wavelength, to 3/2 wavelengths, 4/2 and so on. The lowest possible frequency is

$$f_{min} = c/(2\ L)\ \text{Hz, where } c = 343\ \text{m/s} \qquad ...(6.1)$$

If the excitation is applied as a step function, then the sound pressure will rise from 10% to 90% of its steady-state level within a time

$$T_{rise} = 0.7/BW \qquad ...(6.2)$$

where BW is the width of the resonance curve in Hz at the half power (–.3 dB) level. The SPL will decay to one millionth (–60 dB) of its full level after a time

$$T_{60} = 2.2/BW \qquad ...(6.3)$$

The quality factor or Q of the resonance is

$$Q = n\,f_{min}/BW \qquad ....(6.4)$$

with n = 1, 2, 3, etc.

Example 1

L = 25 ft (7.63 m), then $f_{min} = 343/(2*7.63) = 22.5$ Hz and no resonance below this frequency. The next higher resonance will be at 45 Hz, then 67.5 Hz, 90 Hz, 112.5 Hz and so on.

If we had measured $T_{rise} = 202$ ms at 45 Hz, then from (2) $BW = 0.7/0.202 = 3.5$ Hz and $T_{60} = 2.2/3.5 = 630$ ms from (3).

$Q = 45/3.5 = 12.9$ and if T_{60} stays constant with increasing frequency, then Q increases, for example $Q = 112.5/3.5 = 32.1$

Standing waves in a rectangular, rigid room

In a rectangular room we have six surfaces and the number of possible standing waves is much larger than for the two wall example. The frequencies at which they can occur are calculated from

$$f = (c/2)\,[(l/L)^2 + (w/W)^2 + (h/H)^2]^{1/2} \quad [\text{Hz}] \qquad ...(6.5)$$

$l, w, h = 0, 1, 2, 3$ etc.

A spreadsheet for calculating and plotting room modes and other room parameters discussed here.

At frequencies below the lowest room resonance, the sound pressure will increase at a rate of 12 dB/octave for a closed box speaker that is flat under anechoic conditions, assuming that the room is completely closed and its surfaces are rigid. This case has some significance for the interior of automobiles. Under the same circumstances the sound from a dipole speaker will stay flat.

Domestic listening spaces are seldom completely closed, nor are sheet rock walls rigid, making a prediction of very low frequency in-room response extremely difficult.

Note: Calculations of room modes, though popular, are not practical for predicting optimum speaker placement or listener position. For this one would need to calculate the transfer function between speaker and listener. The transfer function is related to the room modes, but much more difficult to determine. Never-the-less, room mode calculations are often invoked to predict "optimum" room dimensions. They fail to take into account any specifics about speaker placement, source directivity and source type (monopole vs. dipole) that determine which modes are excited, and in combination with the absorption properties of different room surfaces, to which degree these resonances build up. Some people think that by making the room other than rectangular or using curved surfaces, that they can eliminate standing waves. They merely change frequencies, shift their distribution and make their calculation a lot more difficult.

Room modes can be identified by peaks and dips in the frequency response of the acoustic transfer function between speaker and listening position, though only at low frequencies (<150 Hz) where their density is not too high. Such measurements are location dependent and are difficult to interpret as to their audible effect. Listening to a multi-burst test signal at different frequencies gives audible indication of which room locations and frequency regions suffer the greatest degradation in the articulation of bass sounds. With this information in hand it is then possible to identify and electronically equalize the worst offenders in the acoustic transfer function response.

Several room parameters can be calculated that give insight into the general behavior of a closed space.

The number of modes n between zero and a given upper frequency limit f_m can be estimated (H. Kuttruff, Room Acoustics, 1991) from

$$N = (4\,\pi/3)\ V\ (f_m/c)^3 + (\pi/4)\ S\ (f_m/c)^2$$
$$+ (1/8)\ L_e\ (f_m/c) \qquad\qquad ...(6.6)$$

Where

$$V = L\,W\,H \quad [m^3]$$
$$S = 2(L\,W + L\,H + W\,H) \quad [m^2]$$
$$L_e = 4(L + W + H) \quad [m]$$

The number of modes increases very rapidly with frequency and they move ever more closely together. Their average separation at f_m is

$$\delta f = c^3/(4\pi\ V\ f_m^2) \quad [Hz] \qquad\qquad ...(6.7)$$

Example 2

Take a room with $L = 25'$, $W = 16'$ and $H = 9'$ (7.62 m × 4.88 m × 2.74 m), then

$$V = 3600\ ft^3 = 102\ m^3$$
$$S = 1537\ ft^2 = 143\ m^2$$
$$L_e = 200\ ft = 61\ m$$

Below $f_m = 100$ Hz, 200, 300 and 400 Hz the number of modes n and their average separation δf at f_m are respectively

f_m	N	δf
100 Hz	22	3.2 Hz
200 Hz	126	0.8 Hz
300 Hz	375	0.4 Hz
400 Hz	832	0.2 Hz

If we assume that the modes in this room decay at $T_{60} = 630$ ms, then each resonance occupies a 3 dB bandwidth BW = 3.5 Hz from (6.3) above. Somewhere between 100 Hz and 200 Hz the average separation df between modes is 1.2 Hz and thus 3 modes fall within the 3.5 Hz bandwidth resulting from T_{60}. This occurs at $f_s = 157$ Hz as calculated from the simple formula for 3 overlapping modes per BW:

$$f_s = 2000 \, (T_{60}/V)^{1/2} \quad [\text{Hz}] \qquad ...(6.8)$$

The frequency f_s is also called the Schroeder frequency and denotes approximately the boundary between reverberant room behavior above and discrete room modes below.

The sound decay time or reverberation time T_{60} is related to the average wall absorption coefficient α by Sabine's formula

$$T_{60} = 0.163 \, V/(S \, \alpha) \quad [\text{s}] \qquad ...(6.9)$$

$a = 18\%$ for the Example 2 room with $T_{60} = 630$ ms.

It allows evaluating the effect of room modes upon the clarity of sound reproduction.

Reverberation Distance

When we consider radiation in the reverberant frequency range above 149 Hz, the sound at the listening position is composed of the direct sound from the source and the reverberant sound that is more or less uniformly distributed in the room. The direct sound pressure level decreases inversely to distance from the source and will equal the reverberant sound pressure at distance x_r. The 'reverberation distance' x_r (also called 'critical distance') is calculated from

$$x_r = 0.1 \, (\, G \, V/(\pi \, T_{60}) \,)^{1/2} \quad [\text{m}] \qquad ...(6.10)$$

where the directionality gain G is unity for a monopole and $G = 3$ for a dipole radiator. A dipole, thus, has a $3^{1/2} = 1.73$ times larger reverberation distance.

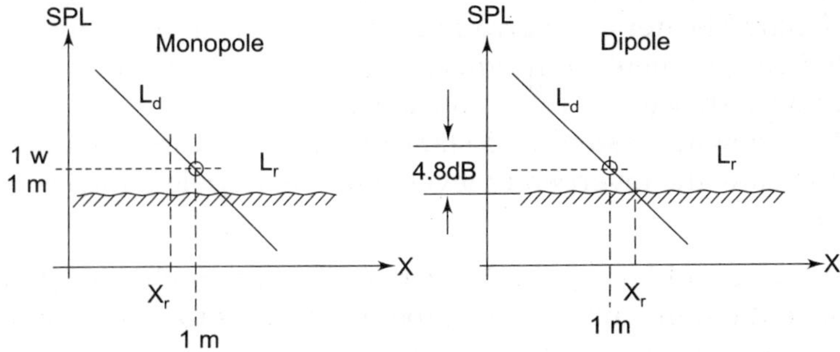

Figure 6.3

You can think of T_{rise} as the time constant of the room. If music or speech varies faster than the time constant, then the room will not respond fully and you hear predominantly the direct sound from the speaker. For 630 ms reverberation time and 200 ms rise time this covers modulation envelopes of a sound down to 1/200 ms = 5 Hz which, in my opinion, is preferable over the 10 Hz envelope rate of a T_{60} = 300 ms room.

In all practical cases the room response time is large compared to the time it takes a reflected sound to reach the listener and therefore reflections will not be masked by the reverberant field. Depending upon the directivity of the source and the proximity of reflecting surfaces and objects specific absorptive or diffusive treatment may become necessary. It should not be overdone, though, because a certain amount of lateral reflection is subjectively desirable to not destroy the impression of a real space.

6.4   LOUDSPEAKER AND LISTENER PLACEMENT

It is often assumed that a study of room acoustics can lead to highly specific loudspeaker and listener placement locations, down to within an inch. Other proponents are not as optimistic and recommend a 1/3rd rule. I have come to the conclusion that real rooms are acoustically far too complex to predict the transmission of sound from speaker to listener, where the sound paths are in three dimensions, have direction and frequency dependent attenuation and diffusion, and can excite the inherent resonance modes of a room to unknown degrees.

From practical experience I recommend the following setups as starting points. They are a dipole or bi-directional loudspeaker, and a monopole or omni-directional speaker. Three room sizes are considered. The 180 ft^2 (17 m^2) room with 8 ft (2.4 m) ceiling would seem like the absolute minimum for quality sound reproduction with the ORION. A 400 ft^2 (37 m^2) or larger room with 10′ (3 m) ceiling should be perfect.

D1-Dipole Setup

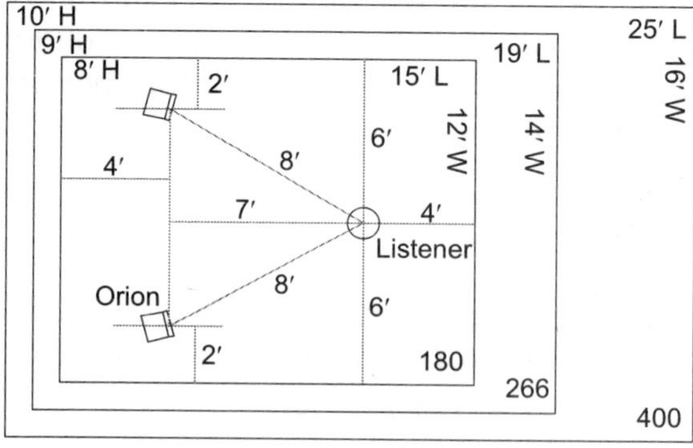

Figure 6.4

ORION separation is 8′. They are slightly towed in. The listener is at the apex of an equilateral triangle. Distance to the wall behind the speakers is 4′, and to the side walls 2′.

The listener is only 4′ from the wall behind, and this might require some heavy curtains and other absorbing material on that wall. As the room gets larger it expands around this triangular setup and especially behind the listener. Sound should just wash by the listener and disappear.

The wall behind the speakers should be diffusive. The rear radiation from a dipole must not be absorbed or it is no longer a dipole. Similarly, the side walls should not absorb sound at the reflection points but diffuse it. A dipole can even be towed in so that the listener sees the radiation null axis in a wall reflection mirror.

D2-Monopole Setup

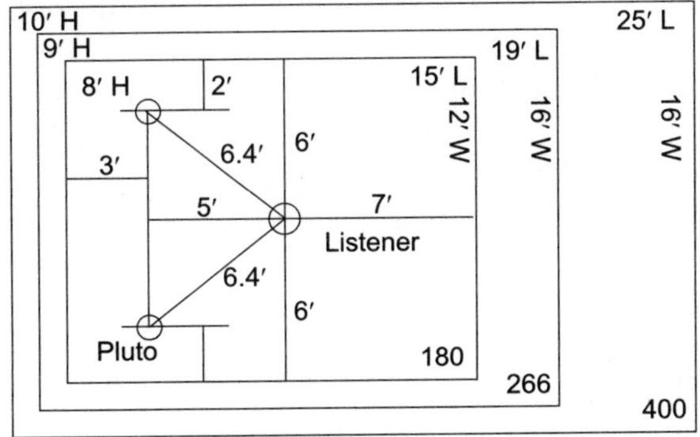

Figure 6.5

PLUTO setup differs from ORION. The listener sits closer to the speakers. The distance to the wall behind the speakers can be slightly less, because of the uniform acoustic illumination of the room. It should not be less than 3′ (<6 ms) to separate reflection from direct sound psycho-acoustically, and to preserve phantom imaging.

Sidewall reflections should be diffused if treated at all. Absorbing them is like turning down the tweeter. Absorbers are not broadband and ineffective below a few hundred Hz. Besides, lateral reflections are important for sound scene recognition.

Again, larger rooms expand around the triangle and increase the space more behind the listener than in front of him.

D3-Pink noise test

Listening to pink noise is a revealing test of electrical and acoustic performance of any system setup. Pink noise must emanate from both speakers simultaneously in dual mono fashion. A tightly confined phantom image should be heard half-way between the two loudspeakers. As you move your head left or right the sound should become brighter sounding and increasingly so with about a 2 inch (5 cm) periodicity as the lateral head displacement is increased in D1. The image also becomes significantly more diffuse and moves towards the nearest speaker. Pink noise should sound neutral and uncolored, though what that exactly means is hard to define. Moving around in the room the character of the noise sound should not change significantly with speakers, holding up even when you leave the room and listen from outside. This is not the case with loudspeakers that have a greatly varying polar response.

Listening to pink noise does not give a reliable indication of system performance at frequencies below 100 Hz and above 10 kHz. Even when pink noise is measured in 1/3rd-octave bands, the response graph is not a reliable indicator of speaker performance and should not be used as the basis for equalization. It seems so obvious that one only needs to have a flat frequency response at the listening position and be done. But, room response equalization is a very complex subject because it deals with sound in three dimensions of space, with time, with frequency, and with a highly evolved auditory stimulus processor between the two ears that is not easily fooled long-term. The response should not be optimized merely at the listening position.

D4-Room analysis

The spreadsheet that was discussed under C above can be used to analyze the three hypothetical rooms and to gain some general insights. Depending upon their structural rigidity, their wall surface textures, floors, floor coverings and objects in different locations, each room will have its own unique acoustic signature. Broadly speaking, a room may sound live or dead. The extremes of this would be an unfurnished room with hard walls versus a cocktail lounge full of overstuffed armchairs and soft leather. Neither one would be suited for sound reproduction. The descriptive parameter is the average absorption coefficient of all surfaces and leakage paths. By definition an open window has a 100% absorption coefficient and if that open window covered 20% of a room's total surface area, then the average absorption coefficient for the room would be 20%. For the 180 ft^2 room example this would be an open window of 169 ft^2 area out of a total surface of 847 ft^2. Since we usually listen with closed windows and very few surfaces have 100% absorption, it takes much more than 169 ft^2 to obtain an overall 20% absorption.

D5-Lively rooms

Table 6.1 D5-Lively Rooms

Floor area ft²	Volume ft³	Absorption %	T60 ms	F-Schroeder Hz	Reverb-dist monopole-m	Reverb-dist dipole-m	Reverb SPL dipole-dB	Direct/Reverb 2.4 m from dipole dB	1st mode Hz
180	1443	20	452	200	0.54	0.93	0.0	-8.2	37.6
266	2398	20	528	168	0.64	1.11	–1.5	–6.7	29.7
400	4007	20	614	140	0.77	1.33	–3.1	–5.1	22.6

D6-Fairly dead rooms

Table 6.2 D6-Fairly dead Rooms

Floor area ft²	Volume ft³	Absorption %	T_{60} ms	F-Schroeder Hz	Reverb-dist monopole-m	Reverb-dist dipole-m	Reverb SPL dipole-dB rel to 20% Absrp	Direct/Reverb 2.4 m from dipole dB	1st mode Hz
180	1443	40	226	141	0.76	1.31	–3.0	–5.3	37.6
266	2398	40	264	118	0.90	1.57	–4.5	–3.7	29.7
400	4007	40	307	99	1.08	1.88	–6.1	–2.1	22.6

The numbers in tables D5 and D6 are for hypothetical rooms and based on a very simple rigid rectangular room model. Though the numbers look precise they should only be taken as trend indicators. Note the relatively narrow range from 99 Hz to 200 Hz covered by the Schroeder frequency for the different rooms and absorptions. Below this frequency specific room modes can dominate, down to the 1st mode. Above that frequency the mode density becomes so high that a room is better described statistically by its reverberation time. For the typical home listening rooms with relatively large objects and different materials in them, reverberation time usually changes with frequency regions and is not as solid a descriptor as for concert halls. Below the first room mode the sound level becomes independent of location in the room and is a function of the lumped mechanical properties of the room. Similar to the modal region the level can be attenuated or amplified depending on wall surface flexures and leaky openings. The room adds to and subtracts from the loudspeaker's direct sound to varying degrees and in a very complex manner over the whole frequency range of the speaker. Thus the tables can only show trends above the Schroeder frequency.

It can be seen in D5 that the reverberant SPL in the 400 ft^2 room is 3.1 dB below that of the 180 ft^2 room and when the absorption is increased to 40% in D6, it drops by another 3 dB for the same direct sound level. Since we judge loudness by the reverberant sound field this means that the volume control setting has to be increased 3.1 dB for the volume in the 400 ft^2 room to be as loud as in the 180 ft^2 room in D5, and by 6.1 dB for the more absorptive room in D6. Still, this is not much of an increase between the small and the large room. It confirms that it can be used in a wide range of room sizes, if volume levels are set for critical listening in the triangle seat and not for sound reinforcement at a large party.

Under D6 the ratio of direct to reverberant sound level is 3 dB better than for the more lively rooms under D5 with half the absorption. These numbers are for the dipole which inherently is 4.8 dB (3x) better than a monopole. But the monopole in D2 is closer to the listener than the dipole in D1. Thus, in all cases the direct-to-reverberant sound ratio for this monopole at 6.4′ (1.92 m) listening distance is only 2.8 dB worse than that for the dipole at 8′ (2.4 m).

Despite the poorer signal-to-reverberant ratio I find more lively rooms preferable for music and voice reproduction. Home Theater installers, though, try to get rooms down to the 200 ms T60 region, which is difficult to accomplish for low frequencies.

Reverberation time of a listening room can be measured rather easily with the Analyzer, but a loud hand clap can tell already whether a room is alive or dead. Rather than special products for acoustic treatment of a room I prefer the normal stuff of life-books, curtains, pictures, rugs, wall hangings, shelves, cabinets, chairs, sofas, etc.-to establish the acoustics of my living spaces.

7

Acoustics and Design

7.1 INTRODUCTION

What is covered:

- Basic acoustic terminology.

- Noise sources, design criteria for different buildings and spaces, assessment of noise levels, and noise control.

- Design issues associated with acoustic performance inside buildings due to internal or external noise sources.

What is not covered:

- Buildings where there are special acoustic constraints e.g. auditoria.

- Factories (and buildings where there is 24 hour work, e.g. hospitals) where it may be important to assess the effect of noise generated on adjacent dwellings.

- Sound systems in buildings. These may be required for emergency warning (e.g. fire alarm), paging system, lecture and conference rooms, sports stadia, railway stations etc.

Examples

- External environment: buildings adjacent to motorways where there may be a need for a sealed building with mechanical ventilation; How noisy can it be before a building cannot be naturally ventilated?

- Internal environment: Office space within factories next to noisy process plant. How can sound levels in offices be made acceptable?

Pumps: Other equipment: Boilers, motors, compressors etc.

Noise in Airflow Systems

Larger diameter ducts - lower air velocity, less noise.

- also need to reduce abrupt transitions to avoid turbulence can add sound absorbent lining or attenuators (silencers).

- beware of possibility of breakout at any airgaps.

    Diffusers - data from manufacturer, or estimate from CIBSE:

    $$PWL = 32 + 13 \log_{10} A + 60 \log_{10} v$$

    where A is the minimum open area in m^2; v is air speed in m/s (e.g. for an air speed of 4 m/s and a 200 mm × 200 mm opening; PWL ~ 50 dB).

External Noise

- Road traffic
    - o calculation of predicted SPL for new roads
    - o measurement or calculation for existing roads
- Aircraft
- Rail
- Industrial sources - in general requires site survey to BS4142
- External equipment and plant

7.4   DESIGN CRITERIA

Regulations

- Noise at Work Regulations - legal duties of employers (and equipment suppliers) to minimize hearing damage.

- Town and Country Planning Regulations - define environmental assessments for any major projects of more than local importance, or projects in sensitive areas.

- Detailed Building Regulations - performance criteria by conforming to design or by measurement - but only for housing.

Houses

For houses, background noise in the house due to external noise sources should be:

- $<$35 dB L_{Aeq} for the period 23:00 to 07:00 in bedrooms;
- $<$40 dB L_{Aeq} for the period 07:00 to 23:00 in living and dining rooms;
- $<$50 dB L_{Aeq} for the period 07:00 to 23:00 in less sensitive rooms.

  The Building Regulations give acceptable constructions and connections for all parts of the building.

Other Buildings

For other buildings, Noise Rating criteria are used:

- NR curves
- recommended noise ratings for spaces
- speech intelligibility (privacy)

  There are no regulations governing acceptable noise levels in offices. BS8233:1987 recommends 40-45dB L_{Aeq} for private offices and small conference rooms, and 45-50dB L_{Aeq} for open-plan offices. This indicates that where external noise levels are in excess of 60dB L_{Aeq} (e.g. from road traffic noise), then a sealed office with mechanical ventilation will be required.

Reverberation Time

Acceptable reverberation times can be specified for rooms for best reception for speech or music. This is primarily of interest to large specialised spaces - auditoria, lecture theatres etc.

7.5 ASSESSMENT OF ROOM SOUND LEVEL

To find the total sound pressure levels in a room: define individual noise sources and their PWL, include the modifying characteristics of the transmission paths (e.g. SRI), apply the acoustic properties of the receiving room (amount of acoustic absorption), and sum.

Outside Noise Environment

This is important, as it has implications for ventilation, and possibly glazing/ constructions e.g. near airports or busy roads. Considerations include:

- External barriers around site - height is critical: note the potential impact on shading;
- Magnitude of noise sources by measurement, or in case of traffic, calculation based on vehicle flow rates, speed, ratio of heavy/light

vehicles, road surface, gradient, distance from road to building, screening correction.

- Distance is important: with vegetation and <4 m reception point, as high as 7 dBA for doubling of distance; with a hard surface or water only 3 dBA for a doubling of distance.

Internal Noise Environment

Consider paths for transmission in buildings.

$$SPL_r = SPL_s - SRI + 10 \log_{10} (S_w/A)$$

where: SPL_r is the sound pressure level in receiving room; SPL_s is the sound pressure level in source room; SRI is the sound reduction index; S_w is area of separating wall; A is total absorption in receiving room (surface area × average absorption coefficient + absorption of furniture/people, m^2).

7.6   CONTROL OF NOISE

In all cases consider (in order) source, transmission path and receiver.

Planning to Control External Noise

- Control of source usually not possible (except by planning constraints).
- Transmission path can be influenced by:
  - location of the building on the site;
  - screening of the site;
  - internal planning of the building;
  - building form and orientation.
- Control at receiver by improving insulation of the building envelope, but the site itself may not be protected, so gardens/public areas may be noisy. The building must be well sealed to give maximum insulation, requiring mechanical ventilation.

Planning to Control Internal Noise

- Reduce noise at source where possible (e.g. acoustic enclosures for noisy machinery).
- Internal planning - ensure that adjacent rooms are compatible in terms of noise sensitivity and noise production.
- Improve room-to-room sound insulation.

Use of Mass

The sound insulation of any single-leaf wall or floor built without gaps depends mainly on its MASS. According to the MASS LAW, there will be an increase in sound insulation of about 5 dB if the mass/unit area is doubled. The insulation also increases by about 6 dB for a doubling of frequency. However, this is only true up to a critical frequency, beyond which there will be a dip in insulation. The critical frequency is about 100 Hz for a one-brick wall, 200 Hz for a half-brick wall. Critical frequencies in the range 100 Hz to 1000 Hz should be avoided.

Use of Isolation

Double leaf walls give good insulation if they are completely decoupled. For example, 2 sheets of plasterboard bonded together will give 30 dB attenuation; this would increase to 50 dB if they were perfectly isolated. In practice, attenuation may vary according to how rigid the link is between the two leaves and the width of the airgap. Absorbent quilt in the airgap improves performance - not because it is a good sound absorber, but because it helps to isolate the two leaves of the partition.

High levels of insulation can be achieved with care and expense - for example, separation of multiplex cinema auditoria of weighted standardised level difference (D_{nT}) of 65 dB to over 70 dB has been achieved by using 2 layers of 15 mm plasterboard on separate studs, a large cavity with 100 mm quilt inlay and careful head, base and edge detailing.

Control of Flanking Transmission

Detailing for houses is given in the Building Regulations. If flanking constructions are not properly specified (and constructed), the flanking transmission can equal or even exceed direct transmission.

Quality of Detailing

Small flaws in construction can lead to large differences in insulation. For example, this may be due to:

- Small airgaps in mortar joints or under skirting boards. For example, an opening of area 0.1 m^2 (SRI of 0dB) in a facade of area 25 m^2 (SRI of 50 dB) reduces the overall SRI value to 24 dB.

- Mechanical bridging of air gaps (nails through floating floors etc).

- Excessive flanking transmission.

7.7 EXAMPLE OF DESIGN CONSIDERATIONS: HOTELS

Main Issues

- Hotels vary in standards, but many are near busy roads or airports, or in city centres.
- The main acoustic issues are noise break-in from outside, pnvacy between rooms, and ventilation noise.

Windows

- Windows in hotels have opening lights even in noisy situations; good weatherstripping and double glazing are essential.
- Balconies can give some protection from noise.

Privacy

- A reasonable standard is attained with separating walls and floors having an SRI of 50 dB.
- Creation of a lobby outside the en-suite bathroom will isolate corridor noise.
- Cross-talk attenuation to bathroom extracts will prevent plumbing sounds being transmitted.
- Single door to corridors should be rated at 35 dB.
- Room televisions should not be fixed directly to room separating walls.

Ventilation Noise

- This should be kept within NR 35 in any hotel and down to NR 25 in good standard bedrooms.
- Connections to outside and chiller plant should be remote from bedrooms, or well-screened and attenuated.

7.8 EXAMPLE OF DESIGN CONSIDERATIONS: OFFICES

Main Issues

- Complaints from office workers arise from intrusive outside noise, high noise levels within offices and poor insulation between cellular offices.
- BS8233: 1987 recommends 40-45 dB L_{Aeq} for private offices and office conference rooms, and 45-50 dB L_{Aeq} for open-plan offices.

- Above a general level of 57 dBA, occupants have to raise their voices to offset the background noise, which further raises internal levels.

Outside Noise Levels

- This can influence the form of the office complex: natural ventilation for a 15 m deep template or natural ventilation plus ventilated core for an 18 m deep template allows in traffic and industrial noise. A deep plan sealed fully mechanically ventilated office building gives a more controlled environment.

- 4/12/6 glazing is usually adequate, but better glazing combinations (6/20/10) or even double windows may be required in exceptional circumstances.

Atria

- Glazing panels act as low frequency absorbers but are otherwise strongly reflective.

- With hard floor and wall surfaces, the space will be reverberant without absorptive panelling on about 25 % of the wall surface. Trees and furniture can help as sound absorbers; water features can help to mask sound.

- There is little data available but two centres monitored had ambient noise levels around NC50 (similar to NR50) due to ventilation plant and continuous escalator operation. Sound levels were very uneven.

Internal Noise Levels

- These can be kept reasonable with a sound-absorbing ceiling, carpet and screened workstations.

- Modern office equipment is much quieter than older equipment (e.g. laser printers are typically 64 dBA at 1 m compared to 83 dBA for mechanical printers).

Privacy

- There is a usually a hierarchy of privacy in offices: senior management has offices with greater privacy and lower ambient noise.

- Privacy between workstations is only typically around 17-20 dBA in open-plan offices - less than required for speech privacy. There is some evidence that resulting interruptions can lead to loss of productivity: improving privacy can improve productivity by 3-10%.

as discussed above for more frequency resolution), and the resulting unit is called decibel or (dB). In sound application the reference value is 20 μPa for sound pressure or 10^{12} W/m² for sound intensity. These values give 0 dB sound level, and represent the human threshold of hearing (the lowest level that can be perceived). The threshold of pain or feeling (the level which causes pain in the ears) is about 120 dB as shown in the Figure 7.2:

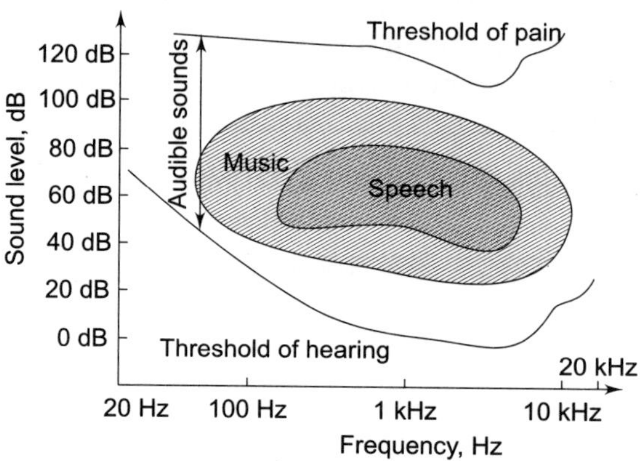

Figure 7.2

Note that the decibel values of sounds cannot be added to obtain the value of the resulting sound. Instead, the combining levels are turned into intensities, added together and then turned back into decibel values.

Sound Absorption Coefficient

Sound absorption coefficient describes the efficiency of the material or the surface to absorb the sound. The ratio of the absorbed sound energy to the incident energy is the sound absorption coefficient. For architectural purposes, sound absorbing materials and constructions can be divided into four types of materials depending on the way the absorption is mainly performed: 1. Turning the sound energy into heat such as fiberglass and carpet. 2. Vibrating with a specific frequency when the sound hits the surface such as lightweight panels and 5/8″ gypsum board. (These materials absorb the sound effectively on a narrow band of frequencies) 3. Turning the sound energy into heat in the neck of the cavities (Helmholtz resonator) such as sound blocks. (This construction has a good absorption on low frequencies) 4. Allowing the sound to go through such as some types of grid systems and lay-in ceiling with sound leakage above it. The most common way to measure sound absorption coefficient is to lay a piece of the material in a reverberant room (a room which has very long reverberation time) then measure the RT so the coefficient can be derived from Sabin equation (the original version of RT calculation). There is a standard that details this procedure. The value of the coefficient for the same material varies with the type of the mounting in the test room. Mounting types that are frequently given in manufacturer's data sheets of the acoustical panels are illustrated in the Figure 7.3:

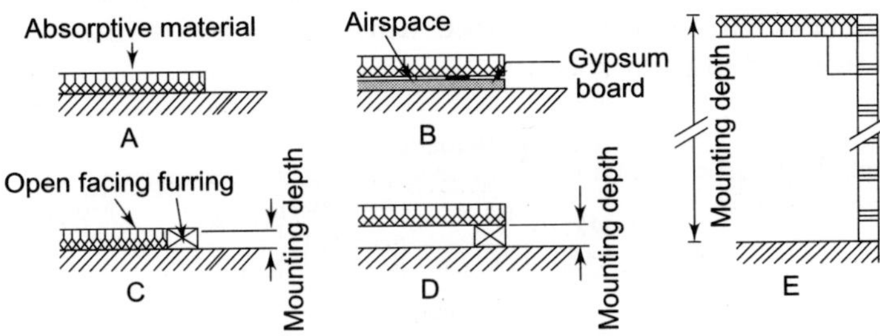

Figure 7.3

Noise reduction coefficient NRC is the arithmetic average of the sound absorption coefficients at 250, 500, 1000, and 2000 Hz, then this average is rounded to the multiples of 0.05.

Diffraction

Diffraction is the change in the direction of the propagation of sound waves passing the edge of the obstacle as illustrated in the Figure 7.4:

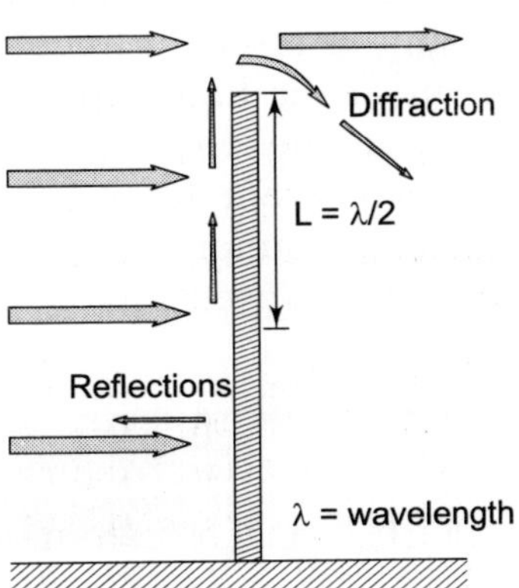

Figure 7.4

Diffraction phenomenon depends significantly on the ratio of the wavelength of the sound to the size of the obstacle. The longer the wavelength the stronger the sound diffraction. Diffraction effect happens to the sound transmitted through openings as well.

7.9 ROOM ACOUSTICS

Physical and Geometrical (Ray) Acoustics

The sound behavior in a room depends significantly on the ratio of the frequency (or the wavelength) of the sound to the size of the room. Therefore, the audible spectrum can be divided into four regions illustrated in the Figure 7.5 (for rectangular room):

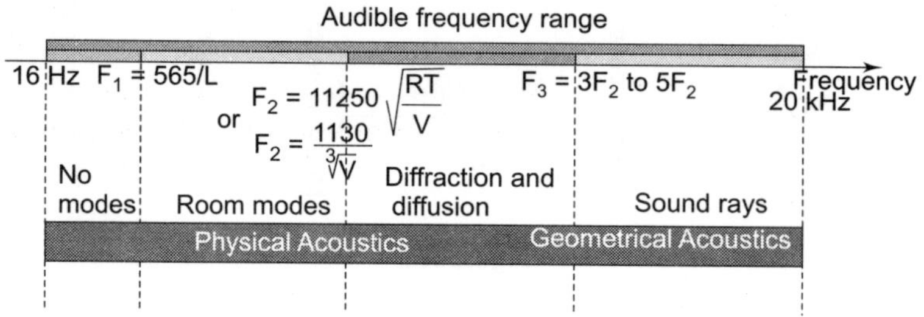

Figure 7.5

Modes

Modes are the resonant frequencies on which the waves interface and form maximums and minimums of sound pressure at different points in the room. The distribution of resonant frequencies over the audible spectrum is not uniform. The spectrum of resonant frequencies is discrete for low frequencies and continuous at the higher frequencies as shown in the following illustration. The larger the room the lower frequency range of the continuous spectrum. The reverberant decay in locations of the maximum sound pressure will be longer, therefore, frequency distortion occurs for sounds with the discrete resonant spectrum.

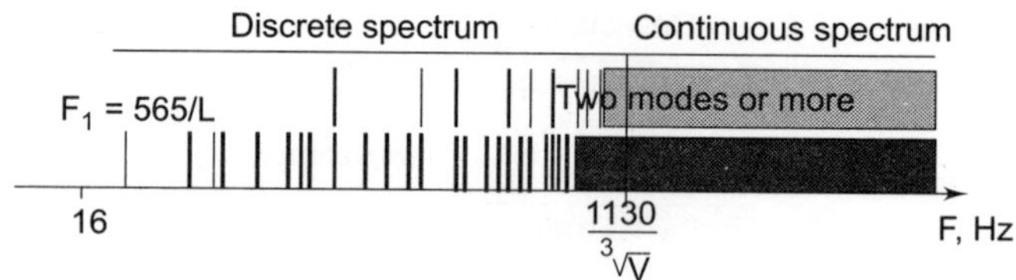

Figure 7.6

The calculation of modes in a rectangular enclosure is simple. The modes become complex and sometimes unpredictable in rooms of different shapes. There are 3 types of modes: axial (two parallel surfaces contribute to the modes), tangential (4 surfaces) and the oblique mode (6 surfaces). The lowest frequency of all modes is for the axial mode, and it can be calculated from

$f = C/2\,L$, where C is the speed of the sound and L is the room length. It is important to note that the modes of the enclosure are weak (in pressure amplitude) when the walls are sound absorbing. To splay the walls (in practice rooms for example) makes modes unpredictable and less organized which, sometimes, can weaken the well defined structure of the maximum and minimum values of the sound pressure.

Sound Diffusion and Diffusers

Sound in an enclosure can be described as a diffused if the intensity of the sound energy is equal in every location of the room, or the sound energy flows equally in every direction. Many different factors can enhance the diffused sound. These factors include geometrical irregularities, absence of focusing surfaces, the distribution of absorptive and reflective elements randomly scattered through the space, and the existence of diffusing objects (furniture) or panels (diffusers).Diffusing panels scatter the sound in all, or in certain directions depending on their type and geometrical dimension. A new type of diffuser is Schroeder diffuser (Quadratic-residue diffusers). Its diffusion characteristics do not depend solely on its geometrical dimensions but also on an array of wells with depths determined by a listed quadratic residue sequence.

Reverberation Time (RT)

Reverberation time is the time required for the sound level in the room to decay 60 dB, or in other words, it is the time needed for a loud sound to be inaudible after turning off the sound source. This concept is shown in the Figure 7.7 and in the example page:

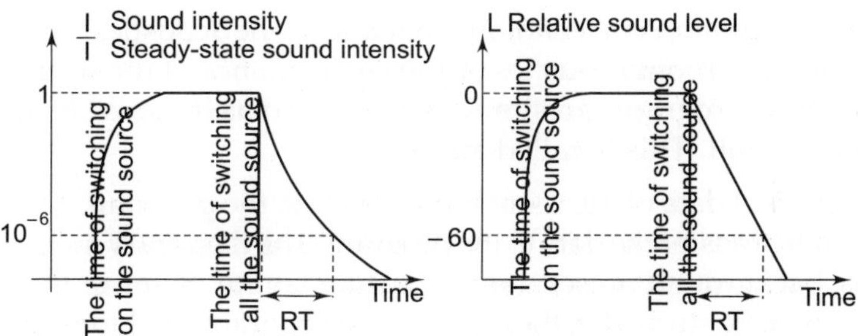

Figure 7.7

The calculation of reverberation time using Sabine or Eyring equations assumes that the sound in the room be diffused. In practice, RT equations are good enough to describe the sound build up and attenuation in the room. In the case where the sound in the room is not diffused enough, such as rooms with good absorption surfaces in some areas, or with an unusual shape (long and narrow, very low ceiling, or many different focusing surfaces), the RT calculation is not accurate. There is Fitzroy equation to correct the RT calculation for rooms with good absorptive surfaces on one (or more) axis of the room. The optimum reverberation time for different rooms depends on the volume of the space, the type of the room, and the frequency of the sound. In general terms, the optimum RT for rooms with speech programs is less then the optimum RT for rooms with music performance.

Reflector: There are few conditions that a smooth-surface panel should meet to be considered a sound reflector. These conditions are illustrated in the following figure:

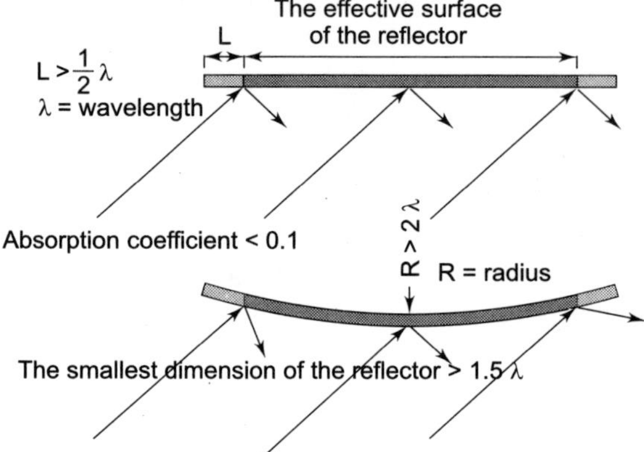

Figure 7.8

Acoustical Simulation

Acoustical simulation is a technique that assists the acoustical consultants in the evaluation of room acoustics or the performance of the sound systems. This acoustical program can simulate the sound as it would be heard after the project is built. This is called auralization.

The physical data of the room is entered into the program. AutoCAD file can be used to transfer the data to the program. The data entry includes surface materials, background noise, and the seating layout. Some of the acoustical factors that can be studied in these acoustical programs include reverberation time, intelligibility, echo, and sound levels over the seating areas.

Noise Control

Noise Paths
Noise in buildings may take many paths. The following figure illustrates the possible paths.

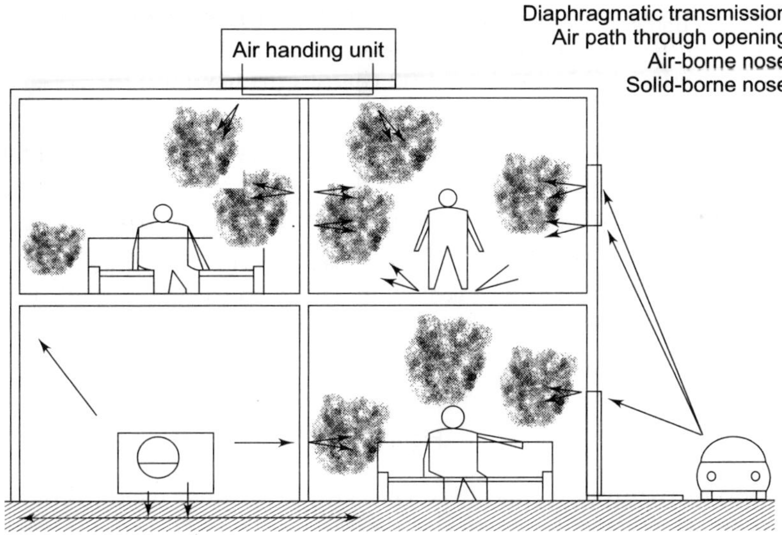

Figure 7.9

Noise Levels

As discussed in the perception of sound levels, the human hearing system has different sensitivities at different frequencies. This means that the perception of noise is not equal at all frequencies. Noise with significant measured levels (in dB) at high or low frequencies will not be as annoying as it would be when its energy is concentrated in the middle frequencies. In other words, the measured noise levels in dB will not reflect the actual human perception of the loudness of the noise. A specific circuit is added to the sound level meter to correct its reading in regard to this concept. This reading is the noise level in dBA. The letter A is added to indicate the correction that was made in the measurement.

The following table 7.3 displays A-weighted sound levels for some common noises:

Table 7.3 A-weighted Sound Levels for some Common Noises

Small office	Large office	Car 65 mph at 25'	Light traffic at 100'	Quiet residential (daytime)	Quiet residential (nighttime)	Sewing machines at 3'
50-55 dBA	60-65 dBA	70-80 dBA	50-60 dBA	40-50 dBA	30-50 dBA	95-100 dBA

TL, STC and IIC

The transmission loss (TL) for a partition and the noise reduction in the room are defined in the following drawing.

transmitted to adjacent rooms is determined by the sound insulating properties of the walls and is not dealt with in this note. For many rooms where the exact behaviour of the sound need not be examined as critically as, for example, in a lecture theatre or a conference hall, sound absorbing materials are used mainly to reduce reverberant noise levels. This paper provides some information to aid in the design of such rooms.

Description of Terms

Some of the more common terms used in this area of acoustics follow.

The *sound absorption coefficient* for a material is the fraction or percentage o incident sound energy that is absorbed by the material. The absorption coefficient depends on the sound frequency and values are usually provided in the literature at the standard frequencies of 125, 250, 500, 1000, 2000 and 4000 Hertz. The *noise reduction coefficient* (NRC) is the average of the values from 250 to 2000 Hz, inclusive, rounded to the nearest 0.05. Absorption coefficients are sometimes expressed as percentages.

The *sound absorption* for a sample of material or an object is measured in sabins or metric sabins. One sabin may be thought of as the absorption of unit area (1 m^2 or 1 ft^2) of a surface that has an absorption coefficient of 1.0 (100 per cent). When areas are measured in square metres, the term metric sabin is used. The absorption for a surface can be found by multiplying its area by its absorption coefficient. Thus for a material with absorption coefficient of 0.5, 10 sq ft has a sound absorption of 5 sabins and 100 m^2 is 50 metric sabins.

The total absorption in a room can be estimated by multiplying the area of each surface by an estimated absorption coefficient for that surface and then adding together the contributions from each surface.

When a source of sound in a 'live' room is turned off, a noticeable time elapses before the noise becomes inaudible. The less sound absorbing material there is in a room the longer the sound takes to die away. The *reverberation time* is defined as the time in seconds required for decaying sound to decrease in level by 60 decibels (dB). Since the sound absorption in the room depends on the frequency of the sounds being considered, so also does the reverberation time.

Reverberation in Rooms

Hard surfaces such as glass, concrete, brick, wood and plaster reflect almost the entire sound incident on them. In contrast with this, soft porous materials such as glass or mineral wool, soft fabrics, clothing, people, and even the air, will absorb sound.

In an extremely 'live' or reverberant room where the surfaces are all good sound reflectors, noise can build up to produce unpleasantly high sound levels and the sound field is constant in level for all distances greater than a few metres from the source. The reverberation times in such rooms can be from 5 to 10 seconds long or more making speech at distances of more than a few metres difficult to understand.

In a reverberant room the two quantities, absorption and reverberation time, T, in seconds, can be related by the equations

$$T = 0.161 \ V/A \qquad \qquad ...(7.1(a))$$

or $$A = 0.161 \ V/T \qquad \qquad ...(7.1(b))$$

Where, V is the volume of the room in cubic metres and A is the total absorption in it in metric sabins (square metres) in the frequency band being considered.

In an extremely absorptive environment the sound level decreases by 6 dB each time the distance from a small source is doubled. There are no reflections, no reverberation and all sound comes directly from the source. This is known as *free field* behaviour.

In typical rooms the behaviour of the sound field can be described approximately with reference to a critical distance D (metre). Within the critical distance the sound field behaves as in the free field case just described, as if there were no reflecting surfaces. This region is called the *near field*.

Beyond the critical distance, in the region called the *reverberant field*, the sound pressure level remains constant. The critical distance is given approximately by

$$D = \sqrt{A/2} \qquad \qquad ...(7.2)$$

where, A is the total absorption in the room in metric sabins.

7.11 REDUCTION OF REVERBERANT SOUND LEVEL

When absorptive material is added to a noisy room, the reduction of sound pressure level in the reverberant field can be predicted from the equation

$$\text{Noise reduction (dB)} = 10 \log_{10}\left(\frac{\text{Absorption after addition}}{\text{Absorption before addition}}\right)...(7.3)$$

For example, doubling the amount of absorption in a room will reduce the reverberant Sound level by 3 dB. To obtain a further 3 dB improvement the amount of absorption must be doubled once again, i.e., four times more than originally existed in the room.

It is important to remember that Equation (7.3) only applies in the reverberant field. Increasing the absorption in a room will reduce noise levels from a machine only for those employees beyond the critical distance. The operator or those close to the machine will be exposed to sound coming directly from it and additional absorptive material will not change the sound levels[*]. Equation (7.2) should be used to estimate where the reverberant field begins.

Types of Absorbing Materials

The absorption coefficients of the many types of sound absorbing materials available are influenced by a number of factors. In general the greater the porosity of a given volume of material is the greater its sound absorbing capability. Thus glass wool is more absorbent than wood chips and nonporous polystyrene foam is a very poor sound absorber. For sheet materials the absorption coefficients vary with mounting technique, e.g., (i) increasing the thickness increases the absorption at all frequencies unless the absorption coefficient is already close to 1.0; (ii) increasing the air gap between the sheet and a solid backing surface increases the absorption at the low frequencies; (iii) covering the sound absorption material with a very lightweight sheet of plastic or a protective layer more than 10 percent open reduces the absorption coefficients slightly at the higher frequencies only.

Absorptive materials are often applied in patches or strips of material or suspended as rectangular panels or other forms. These suspended shapes are known as unit absorbers. It is quite common for materials tested in a laboratory reverberation chamber as a small sample or collection of small samples to exhibit absorption coefficients greater than 1.0. This is a real effect caused by diffraction. In such cases the measured absorption coefficients are very strongly influenced by the size of the patches or samples and the spacing between them. Test reports, usually available from manufactures, should he consulted to determine the appropriate coefficients for any proposed installation.

Many common materials, drapes, carpets and furniture also absorb sound to varying degrees. Table 7.4 illustrates some of the foregoing points with representative values of absorption coefficients for a variety of materials.

---

[*] At exception to this occurs when the noisy machine and its operator are situated close to walls or other reflecting surfaces, here the use of sound absorptive material correctly placed on these surfaces can provide some noise reduction.

Table 7.4 Representative Values of Absorption Coefficients for Some Common Acoustical Materials

Sample	1/3 Octave Centre Frequency (Hz)						NRC
	125	250	500	1000	2000	4000	
Glass fibre on hard surface							
(i) 25 mm (1") thick	0.06	0.25	0.55	0.90	0.90	0.09	0.65
(ii) 50 mm (2") thick	0.02	0.50	0.08	0.95	0.95	0.95	0.80
(iii) 75 mm (3") thick	0.30	0.80	0.95	0.95	1.00	1.00	0.95
(iv) 100 mm (4") thick	0.40	0.90	1.00	1.00	1.00	1.00	1.00
Glass fibre, 50 mm (2") thick							
(i) 25 mm (1") air space behind	0.30	0.70	0.90	1.00	1.00	1.00	0.90
(ii) 25 mm (1") thick on 300 m (12") air space	0.50	0.90	0.80	1.00	1.00	1.00	0.95
25 mm (1") glass fibre covered							
(i) by 5% open perforated layer	0.05	0.25	0.50	0.80	0.25	0.15	0.45
(ii) 25 mm (1") glass fibre covered by 10% open perforated layer	0.05	0.25	0.50	0.80	0.60	0.40	0.45
Wood wool board 25 mm (1") thick							
(i) No air space	0.05	0.15	0.30	0.55	0.65	0.75	0.40
(ii) 25 mm (1") air space	0.10	0.25	0.35	0.55	0.65	0.75	0.45
(iii) 300 mm (12") air space	0.45	0.50	0.45	0.60	0.70	0.80	0.55
Mineral fibre ceiling tiles							
(i) No air space	0.05	0.15	0.50	0.75	0.65	0.50	0.50
(ii) 300 mm (12") air space	0.03	0.40	0.55	0.80	0.70	0.55	0.60
Typical carpet	0.02	0.06	0.15	0.40	0.60	0.65	0.30
Lightweight drapes	0.03	0.05	0.10	0.20	0.25	0.35	0.15
Heavy weight drapes	0.13	0.35	0.60	0.80	0.75	0.75	0.65
Audience in upholstered seats per unit floor area	0.39	0.57	0.08	0.90	0.90	0.85	0.80
Air absorption Relative humidity 50% (metric sabins per 100 m3)	–	–	–	0.30	0.90	0.40	–

Amount of Absorbing Material Required

The types of rooms that can be treated using these equations as a basis for calculation are listed in Table 7.5 together with recommended average reverberation times. These times are meant as a guide and need not be achieved exactly.

Table 7.5 Recommended Average Reverberation Times in Seconds for Spaces which can be Acoustically Treated using this Approach

Space	Average Reverberation Time (seconds)
Classrooms, museum, libraries, restaurants, offices, hotel, apartment and office corridors	0.75 or less
Gymnasiums, machine rooms, computer rooms, swimming pools, airport lobbies, department stores, general purpose assembly rooms, conference rooms, cafeterias	1.0 or less
Sports arenas	2.0 or less

The acoustical treatment of the types of rooms listed is quite straightforward. From the recommended reverberation time and the room volume, the required absorption in sabins is calculated using Equation (7.1(b)). The appropriate choice of type and location of material is then made considering thermal, mechanical and other physical requirements such as condensation control. The area of each material is multiplied by its noise reduction coefficient value to obtain its contribution to the total absorption; the total is compared with the design requirement and any necessary adjustments made. The reverberation time should then be calculated for each frequency band using the appropriate absorption coefficients to ensure reasonable agreement with the recommended average value. It is quite usual for the reverberation time at the lower frequencies to be as much as twice as long as those at the middle frequencies. If the variations with frequency are excessive, change of materials or methods of mounting may be necessary.

In many of the rooms listed in Table 7.5 it will be quite adequate to install a good quality acoustical ceiling, a carpet or the equivalent. This is the case, for example, in hotel or apartment corridors. In small offices and dwelling rooms, the normal furnishings and occupants themselves usually provide enough absorption. In general, however, for best performance acoustical materials should be distributed over as many of the room surfaces as possible. An example of poor application would be an acoustical ceiling in a very high room with very little other absorptive material present. Sound would then be quite free to circulate in the horizontal plane and the room would be more live than calculation would suggest.

In a few of the cases listed in Table 7.5, particularly where speech intelligibility is important, it is necessary to exercise some care in the positioning of the sound absorbing material. For example, in conference room the ceiling immediately above the table should be hard and sound reflecting. The sound absorbing materials should he restricted to the floor, walls and ceiling perimeter. Similarly, in a classroom where the teacher addresses the students from one end of the room, it is good practice to make the centre portion of the ceiling sound reflecting to provide useful voice reflections. In such cases it is probably advisable to obtain professional help or else consult a good textbook.

It is important during the design phase to include the absorptive properties of the normal contents of the space, i.e., people, merchandise, essential furnishings and sometimes air. This would be important in a sports arena, for example. In a department store the absorption provided by the merchandise will vary widely from place to place, but as a rough guide the absorption values for a typical carpet may be assumed.

Example: Determine the necessary acoustical treatment for a gymnasium with a volume of 2720 m^3 and a ceiling area of 340 m^2.

From Table 7.5 the recommended reverberation time is 1 second and from Equation 7.1(b) the required absorption is 438 metric sabins. The sound absorbing material can be applied on both ceiling and walls and should be able to withstand a certain amount of impact. Slotted sound absorbing concrete blocks arc extremely useful in gymnasiums, swimming pools, machine shops and similar places since they can fulfil structural requirements as well as providing sound absorption. The NRC for these blocks is given in Table 7.2 as 0.6.

Suppose, in this case, a spray-on cellulose fibre is to be applied directly to a rigid ceiling (Table 7.5, NRC = 0.55). The resulting ceiling will contribute 187 m sabins. This leaves 251 m sabins to be provided by the slotted sound absorbing blocks. i.e., an area of 418 m^2. The calculated reverberation time at 125 Hz is 1.4 seconds and at 4000 Hz it would fall to 0.93 sec. This is quite acceptable. (The contribution of the air is not important in this case.)

7.12 CONCLUSIONS

This method of approach has definite limitations, but it may he applied without serious error to the types of rooms listed here. The design of special purpose rooms such as auditoriums, theatres, and recording studios and so on is best left in qualified hands.

8 QUESTIONS AND ANSWERS

QUESTIONS AND ANSWERS

1.   Does adding sound-absorbing material to the surface of a wall or a floor significantly increase the sound insulation?

The addition of a typical sound-absorbing material to the surface of a wall may change the high frequency transmission losses. This is where common sound-absorbing materials are most effective. However, the sound transmission class does not usually change, or only changes by one or two decibels. This is because the low frequency transmission is not changed by the addition of typical absorptive materials on walls. The addition of a carpet to a floor, will certainly increase impact insulation ratings, but usually does not change the sound transmission class. Many people believe that adding sound-absorptive material to walls or doors will improve sound isolation between apartments. On the receiving room side, if enough absorptive material is added, the room will be less reverberant and intrusive noise levels will be slightly reduced; the reduction can be as much as 3 dB if the amount of absorption in the room is doubled. The room will also feel 'dead' acoustically, so there is a perception of some improvement. In a normally 'dead' room like a bedroom, it is very difficult to make significant changes to the amount of absorption because it is already so absorptive. Careful measurements have shown that adding absorption to reduce sound transmission is not the most effective way of dealing with this problem.

2.   Is there any information on how much extra it costs to provide good sound insulation and low background noise in a building?

Most cost studies have looked at the incremental cost of materials needed to improve the STC ratings of walls. In general, the cost of the extra materials to improve a specific type of wall is relatively small. It may involve only the addition of some extra wallboard, some sound-absorbing material or the use of a larger airspace.

Other costs, which have not been studied, are associated with the inspection of drawings, site inspections and measurements by a qualified acoustical consultant. One consultant estimated that these costs amounted to about $1000 per apartment unit, a small fraction of the total cost.

3.   What can be done to reduce sound transmission by way of the plenum space in an office building?

In typical office buildings the sound travels through the ceiling material, through any gaps in the ceiling and into the plenum. It then travels freely in the plenum and down by similar paths into adjacent offices. Typical ceiling board materials provide limited sound attenuation, especially if there are ventilation and other openings. To improve the situation, one tries to attenuate the sound passing through the ceiling or in the plenum path as much as possible. Closing off the gap above the partition with blocking panels of materials such as gypsum wallboard or mass loaded vinyl, will give significant improvements. Good results have been obtained recently by stuffing the gap with a pile of glass fibre batts about 400 mm wide. All leaks should be reduced as much as possible, especially those at the junction of the partition and the ceiling. Since the plenum is usually used for ventilation, any air flow paths that might increase sound transmission should have adequate sound attenuation, for example, lined ducts should be used.

4.   If apartment doors have improved sound insulation, as you suggest, does this not mean that fire alarms are likely to be less audible?

Yes, if the alarms are in the corridor. However, even in apartments where doors are not particularly good, fire alarms may not be audible if there are too many doors between the alarm and the occupants. It is preferable to have fire alarms inside apartments, with electrical connections so that all alarms are triggered by one. When selecting the location of fire alarms in a home or an apartment, it is necessary to know the sound power of the device, so that reasonably accurate predictions of its effectiveness can be made.

5.   What is involved in making a measurement of sound transmission loss in a building?

A loudspeaker and amplifier are used to generate loud random noise in a room on one side of the partition under test. For at least 16 frequencies, starting at 125 Hz and ending at 4000 Hz, the average sound pressure level is measured in this source room and the room on the other side of the partition. The difference between these two sets of numbers is called the noise reduction. The larger the area of the partition separating the two rooms, the more sound energy will pass through it. The more absorptive the receiving room, the lower the sound pressure level in that room will be. The noise reduction values are therefore adjusted to compensate for these two factors to give the sound transmission loss; this is a measure of the sound power transmitted per unit area and is independent of the area of the partition and the receiving room properties. The transmission losses are used to calculate the sound transmission class.

ASTM E597 describes a quicker procedure, which gives a number very close to the field sound transmission class. It provides a convenient way of rapidly testing several or all of the party walls and floors in an apartment without the expense of the full test. Ratings from E597 are usually within about 2 dB of the FSTC rating for the partition.

6. What is the difference between noise isolation class and field sound transmission class?

If noise reduction values are treated as though they were transmission loss values, and the sound transmission loss contour is fitted in the usual fashion, then the number so derived is called the noise isolation class. No adjustments are made for partition area or room absorption and no steps are taken to eliminate flanking paths. The noise isolation class therefore characterizes the sound transmission between two rooms, taking into account all flanking paths and the physical conditions at the time of measurement.

7. How does the field sound transmission class relate to the value measured in the laboratory?

There is some controversy over this issue. If there is significant flanking transmission in a building, the apparent sound transmission class will be lowered. One can expect to get ratings as much as 5 dB lower in the field than those obtained in the laboratory, but with good construction practice, values almost as high as those in the laboratory can be achieved.

8. If the sound insulation were measured for a large number of party walls in an apartment, would the FSTC ratings vary significantly?

This is an area where not enough is known. A certain amount of variation is associated with the test procedure itself, but this is small relative to the amount of variation associated with the building. This will be determined by the degree of care used to make sure that partitions and floors are identical and constructed properly and that there is no flanking transmission. However, if a field test is performed and the field sound transmission class is found to be 5 dB lower than the expected value, then something is wrong.

9.   Does plastic piping increase or reduce plumbing noise?

The type of piping used in plumbing systems is not the major factor controlling the noise generated in a building. There are differences between the various materials and in general, the heavier the pipe, the less sound is radiated. The noise radiated from pipes is usually negligible compared to that radiated from the partitions and the structure. It is much more important to ensure that all resilient supports for the plumbing system are properly installed and that quiet fixtures are used where possible.

10.   What kinds of resilient materials are used below floating floors?

A floating floor system is designed to have a fundamental resonance far enough below the frequencies of interest that vibration transmission is significantly reduced for frequencies higher than this resonance. The mass of the floating layer and the stiffness of the resilient supports determine this frequency. Materials used as resilient supports include rubber or neoprene pads, cork pads, some special composite materials such as glass fibre/ neoprene and, in some critical applications, steel coil springs. In some cases, the whole subfloor is covered with a blanket of mineral wool, which provides sound absorption as well as resilient support.

11.   Can suspended absorbing panels be used to control reverberation in spaces such as a school gymnasium, and how should they be designed and installed for maximum effect?

Suspended absorbing panels or unit absorbers are an effective and practical solution to controlling reverberation in spaces such as industrial locations and school gymnasia. Although manufactured unit absorbers of various shapes are available, one can construct units from standard 0.61 by 1.22 m (2 by 4 ft) rigid absorbing panels such as those used in some suspended ceilings. Placing two 2.5 cm thick panels back to back in a suitable frame will produce a unit absorber with more than double the sound absorption expected from tests on larger panels of the same total thickness of material (5 cm) measured against a rigid backing. This is due to the small size of the panel and to the fact that both sides of the panel are exposed to the sound field. The low frequency performance of the unit can be increased by including an air space between the two absorbing layers, or by using thicker layers of absorbing material.

It is generally more satisfactory to hang panels vertically so that both sides of the panel are fully exposed to the sound field. The panels should be distributed evenly throughout the room. When panels are hung horizontally, there will be decreased absorption at medium and higher frequencies as the panel is moved closer to the ceiling. With horizontal panels, increased absorption can be produced over particular narrow low frequency ranges, but these increases are not large and will be associated with a loss of absorption at medium and high frequencies. When panels are installed too close together, they tend to shadow each other and the absorption per panel will decrease. Results from one set of tests for 0.61 by 1.22 m panels suggest that this effect becomes significant at densities of greater than about one panel per three square metres of ceiling. Of course, as more panels are added there will always be an increase in the total sound absorption, but the average absorption per panel may decrease.

12. Can carpeting be used on walls as a physically rugged, but acoustically effective sound-absorbing material?

This is probably not the most cost-effective solution and in some cases may not be permitted by fire regulations. In general, carpets are only good sound absorbers at high frequencies, and not very effective at lower frequencies, due to their not being very thick. Porous sound-absorbing materials must approach 1/4 wavelength in thickness to be optimum absorbers. At a frequency of 100 Hz, typical of many bass sounds in music, the wavelength of sound is 3.4m. Thus it is nearly impossible to have an optimum porous absorber at this frequency. However at higher frequencies, where the wavelength is shorter, thin porous layers such as carpets are more effective sound absorbers. The actual mid-frequency absorption of carpets can vary considerably. Where the carpet is porous to airflow, then thicker and denser carpets will tend to be better absorbers. In some cases, carpets that are rubber-backed or that have a sealed backing on a foam pad can have greater mid-frequency absorption because of membrane absorber effects.

Other absorbing materials can be protected by covering them with wooden slats, perforated metal, expanded metal, or other such constructions. One should leave at least fifty percent of the porous absorbing material exposed to sound, to avoid reducing its effectiveness. When the percentage of exposed surface is much less than this, resonant tuned cavity absorption effects can occur; increased absorption will be found at particular frequencies, while overall absorption will decrease.

13. How should one select a duct silencer?

One must consider three factors: the acoustical insertion loss or attenuation, the self noise or flow noise produced by the silencer, and the static pressure drop induced by the silencer. When a muffler is found with a suitable static pressure drop, the insertion loss in decibels must be large enough to reduce the fan noise levels below specified criteria for the rooms that the fan is supplying. Muffler insertion loss is usually specified in octave band frequency ranges. Typically, silencers are much less effective at low frequencies and adequate performance in this range is particularly important. One should obtain the manufacturer's insertion loss data measured at approximately the same flow velocity as will be used in the planned installation, because insertion loss values tend to decrease with increased air flow velocity.

It is also important to verify that self noise of the silencer (the noise produced by air flow through the silencer) is not substantially greater than the attenuated fan noise. The self noise will vary considerably with air flow velocity, so the noise generated at the velocity planned for the installation must be known. If the self noise exceeds the attenuated fan noise, then an optimum silencer has not been chosen. Another silencer should be selected, with a little less insertion loss and less self noise. For example, many mufflers splitters (absorbent structures in the middle of the muffler); these tend to increase the insertion loss but also to increase the self noise and the static pressure drop of the muffler. Thus, where the maximum possible insertion loss is not required, there are choices to optimize the effectiveness of the silencer.

14. Does painting a ceiling tile reduce its sound absorption?

Ceiling tiles absorb sound because they are porous to the movement of air associated with sound. Painting a ceiling tile could completely seal the surface and drastically reduce its effectiveness as a sound absorber. It may be possible to add one very thin coat of non-bridging latex paint without severe degradation of the absorption of the tiles. Tiles that have rough textured surfaces will be less severely degraded by such a coat of paint, as the surface is less likely to be sealed. The method of application is also of some importance. Spray applications will drive the paint into the pores of the tiles and vigorous brushing could have the same effect. Light application with a roller should give the best results.

15. Auditorium acoustics seems to be more an art, or perhaps a black art, than a science. Are there newer, more scientific, approaches to assessing the acoustical problems of auditoria?

The question of auditorium acoustics is a little beyond the scope of these seminars, but there certainly have been considerable developments in this field. Reverberation time is no longer the sole objective measure for assessing a hall. A number of newer acoustical measures relate to subjective assessments of halls. NRC has developed efficient methods for measuring these quantities, and has assisted clients by evaluating various auditoria.

16. Does the stud spacing in a wall affect the sound transmission class?

From theoretical considerations, we expect that the STC may be slightly reduced by reducing the spacing between studs or by decreasing the space between the nails or screws fastening the gypsum board to the studs. Only a few laboratory tests of these effects are available, and no clear conclusions can be drawn from them. For typical spacing of studs and fasteners, the STC is unlikely to vary by as much as 2 dB.

17.   Are all resilient channels and all lightweight steel studs the same?

How much vibrational energy is transferred through resilient steel channels or steel studs depends on their stiffness. This depends not only on the material (usually 24 gauge sheet steel) but also on how they are formed. Independent test data for reliable comparison of the effectiveness of various products are not available.

18.   Is there a significant difference between solid wood floor joists, truss joists, or wooden trusses? Is it better to use separate joists to support the ceiling in a floor system?

For acoustical purposes, there is little difference among common joist types. They must be quite stiff to support the floor, and hence effectively transmit vibrations from the floor above to the ceiling below (or vice versa). Using resilient channels to support the ceiling reduces this problem, and using separate joists to support the ceiling is even better. In actual buildings, eliminating this path for vibration transfer may give little benefit because of vibration transfer between the floor and the supporting walls.

19.   How much sound-insulating material is optimum in a wall or a floor? Are the fit of the insulation in the cavity and the density or absorption coefficient of the sound-insulating material significant?

The highest STC should be obtained when the cavity is almost full, but the difference between two-thirds full and completely full is unlikely to change the STC of typical partitions by more than one decibel. The absorptive material must not be packed too tightly into the cavity, as this can provide an additional path for vibration transfer. A porous, acoustically absorbing material is needed, but available test data have not clearly established strong dependence of the transmission loss on the absorption coefficient or density.

20.   Is there a preferred way of attaching multiple layers of wallboard together? Is glue better than screws? Is there any benefit to having unbalanced construction (i.e., more sheets of wallboard on one side than the other) in walls or floors?

Multiple layers should be attached together so that the intervening airspace is as small as possible (preferably much less than 1 mm) to avoid problems with the mass--air--mass resonance. The difference between using screws or spots of glue appears to be small, and with these methods of fastening, the distribution of sheets of wallboard between the two faces has little effect on sound transmission. There is a slight benefit from using layers with different coincidence frequencies, but this generally affects only high

REFERENCES

Book

Noise Control in Buildings: A Guide for Architects and Engineers by Cyril M. Harris, McGraw-Hill (Sep 1993).

Architectural Acoustics by M. David Egan, J. Ross Pub (Jan 2007).

Noise Control Manual for Residential Buildings by David Harris, McGraw-Hill Professional; 1 edition (Jul 1997).

Acoustics and Sound Insulation (Details Practice) by Eckard Mommertz, Birkhäuser Architecture; 1 edition (Mar 2009).

Engineering Noise Control: Theory and Practice by David A. Bies and Colin H. Hansen, CRC Press, 4 edition (Aug 2009).

Sound Insulation: Theory into Practice by Carl Hopkins, A Butterworth Heinemann Title, 1 edition (Nov 2008).

Master Handbook of Acoustics by F. Alton Everest and Ken C Pohlmann, Tab Electronics, 5 edition (Jul 209).

Noise and Vibration Control Engineering: Principles and Applications by István L. Vér and Leo L. Beranek, Wiley; 2 edition (Nov 2005).

Foundations of Engineering Acoustics by Frank J. Fahy, Academic Press; 1 edition (Oct 2000).

Fundamentals of Noise and Vibration Analysis for Engineers by M. P. Norton, Cambridge University Press; 2 edition (Nov 2003).

Practical Guide to Noise and Vibration Control for HVAC Systems SI by Mark E Schaffer, American Society of Heating, Refrigerating and Air-Conditioning Engineers; 2 edition (Jan 2011).

Handbook of Noise and Vibration Control by Malcolm J. Crocker, Wiley; 1 edition (Oct 2007).

Fundamentals of Acoustics by Lawrence E. Kinsler, Austin R. Frey, Alan B. Coppens and James V. Sanders, Wiley; 4 edition (Dec 1999)

Construction Vibrations by Charles H. Dowding, C H Dowding; 2 edition (Dec 2005).

Room Acoustics by Heinrich Kuttruff, CRC Press; 5 edition (Jun, 2009).

Architectural Acoustics (Applications of Modern Acoustics) by Marshall Long, Academic Press; 1 edition (Jun 2005).

Acoustic Absorbers and Diffusers: Theory, Design and Application by Trevor J. Cox and Peter D'Antonio, CRC Press; 2 edition (Mar 2009).

Acoustics: Sound Fields and Transducers by Leo L. Beranek and Tim Mellow, Academic Press; 1 edition (Oct 2012).

Papers

Mahavir Singh, Omkar Sharma and V. Mohanan, "Factors Affecting Sound Transmission Loss," Journal of the Acoustical Society of India, 35(1), pp. 40-50, Jan. 2008.

Mahavir Singh, Omkar Sharma and V. Mohanan, "Sound Transmission Loss through Lightweight Wall Structures," Journal of the Acoustical Society of India, 35(2), pp. 94-99, Apr. 2008.

Mahavir Singh, Omkar Sharma and Dharam Pal Singh, "Some Physical Parameters Affecting Sound Transmission Loss," Proc. of National Symposium on Acoustics (NSA2008), Visakhapatnam, pp. 353-358, Dec. 2008.

Mahavir Singh, Omkar Sharma and Dharam Pal Singh, "Reduction of Sound Transmission through Lightweight Gypsum Board Walls," Proc. of 7th International Conference on Advances in Metrology (AdMet-2009), New Delhi, pp. 120-122, Feb. 2009.

Mahavir Singh, Omkar Sharma and Dharam Pal Singh, "Reducing Noise in Your Apartment," Proc. of 7th International Conference on Advances in Metrology (AdMet-2009), New Delhi, pp. 123-125, Feb. 2009.

Mahavir Singh, Omkar Sharma and Dharam Pal Singh, "Sound Transmission through Lightweight Single and Double Leaf Partitions," Journal of the Acoustical Society of India, 36(2), pp. 96-101, Apr. 2009.

Mahavir Singh, Naveen Garg, Omkar Sharma and V. Mohanan, "Akal Parat wale Panlon mai Dwani Sancharan Hras ka Parikalan," Bhartiya Vaigyanik Evam Audyogik Anusandhan Patrika, 17(1), pp. 60-63, Jun. 2009 (in Hindi).

Mahavir Singh, Omkar Sharma and Dharam Pal Singh, "Sound Transmission through Lightweight Concrete Walls," Proc. of 16th International Congress on Sound and Vibration (ICSV16), Kraków, pp. 1-10, Jul. 2009.

Mahavir Singh, Omkar Sharma and Dharam Pal Singh, "Sound Transmission through Both Lightweight and Masonry Single and Double Leaf Walls," Proc. of 2nd National Conference on Innovations in Indian Science, Engineering & Technology (NCISET 2009), New Delhi, pp. 1-7, Jul. 2009.

Mahavir Singh, Omkar Sharma and Dharam Pal Singh, "Sound Transmission Loss of Lightweight Walls," Proc. of 10th Western Pacific Acoustics Conference (WESPAC X), Beijing, pp. 1-10, Sep. 2009.

Mahavir Singh, Omkar Sharma and Dharam Pal Singh, "Sound Transmission through Lightweight Partitions," Proc. of ACOUSTICS High Tatras 2009 "34th International Acoustical Conference – EAA Symposium", Slovakia, pp. 1-10, Sep. 2009.

Mahavir Singh, Naveen Garg, Omkar Sharma and V. Mohanan, "Bilateral Comparison of Pressure Sensitivities of Laboratory Standard Microphones between NPLI and DPLA-DFM," Journal of the Acoustical Society of India, 36(4), pp. 144-151, Oct. 2009.

Mahavir Singh, Omkar Sharma and Dharam Pal Singh, "Sound Transmission through Wall Elements," Proc. of National Symposium on Acoustics (NSA2009), Hyderabad, pp. 1-6, Nov. 2009.

Mahavir Singh, Dharam Pal Singh and Omkar Sharma, "Airborne Sound Transmission in Lightweight Wall Design Structures for Residential and Commercial Buildings," Proc. of the 20th International Congress on Acoustics (ICA-2010), Sydney, pp. 1-6, Aug. 2010.

Mahavir Singh, Omkar Sharma and Dharam Pal Singh, "Vibration in Dwelling Units from Road Traffic," Proc. of National Symposium on Acoustics (NSA2010), Rishikesh, pp. 66-72, Nov. 2010.

Mahavir Singh, Omkar Sharma and Dharam Pal Singh, "Controlling Reverberant Noise in Rooms using Sound Absorptive Materials," Proc. of National Symposium on Acoustics (NSA2010), Rishikesh, pp. 73-78, Nov. 2010.

Mahavir Singh, Omkar Sharma and Dharam Pal Singh, "Room Acoustics - Design for Internal Acoustic Quality," Journal of the Acoustical Society of India, 38(1), pp. 3-10, Jan. 2011.

Mahavir Singh, "Sound Transmission through Building Elements," Proc. of Indo-US Workshop on Nanosonic & Ultrasound (IUWONU2011) and International Conference on Nanotechnology & Ultrasound (ICNU2011), Trichy, pp. 52-70, Jan. 2011.

Mahavir Singh, Omkar Sharma and Dharam Pal Singh, "Experimental Investigations of the Sound Transmission Loss of Window Panels," Proc. of 1st National Conference on Advances in Metrology (AdMet-2011), Banglore, pp. 1-6, Feb. 2011.

Mahavir Singh, Omkar Sharma and Dharam Pal Singh, "Acoustic Comfort – Noise Control for Buildings," New Dimensions of Physics of Proc. of National Conference on Ultrasonics (NCU-2011), Jhansi, pp. 1-14, Mar. 2011.

Mahavir Singh, Omkar Sharma and Dharam Pal Singh, "Sound Transmission through Cavity Walls Constructed from Gypsum Board," GESTS Int'l Trans. Acoustics Science and Engr., 15 (3), pp. 107-112, Mar. 2011.

Mahavir Singh, Omkar Sharma and Dharam Pal Singh, "Experimental Analysis of the Sound Transmission Loss of Single, Double and Triple Glazing Window for Traffic Noise Reduction, Journal of the Acoustical Society of India, 38(3), pp. 106-113, Jul. 2011.

Mahavir Singh, Dharam Pal Singh and Omkar Sharma, "Sound Insulation Performance of Double-leaf Structures," Acoustic Waves of Proc. of National Symposium on Acoustics (NSA2011), Bundelkhand, pp. 15-24, Nov. 2011 (Received a Best Paper Award).

Mahavir Singh, Omkar Sharma and Dharam Pal Singh, "Sound Transmission Loss of a New Designed Lightweight Partition," Acoustic Waves of Proc. of National Symposium on Acoustics (NSA2011), Bundelkhand, pp. 62-77, Nov. 2011.

Mahavir Singh, Omkar Sharma and Dharam Pal Singh, "Vibration in Residential Environments due to Road Transportation," Journal of the Sound, Vibration and Hardness, 1(1), pp. 00-00, Dec. 2011.

Mahavir Singh, Omkar Sharma and Dharam Pal Singh, "Experimental and Theoretical Evaluation of Sound Transmission through Double Window Panels," GESTS Int'l Trans. Acoustics Science and Engr., 15 (3), pp. 00-00, Jul. 2011.

Mahavir Singh, Omkar Sharma and Dharam Pal Singh, "Some Practical Aspects of Absorption Measurements in Reverberation Rooms," GESTS Int'l Trans. Acoustics Science and Engr., 15 (3), pp. 00-00, Sep. 2011.

Mahavir Singh, Omkar Sharma and Dharam Pal Singh, "Theoretical and Experimental Investigation for Sound Transmission through Multi-layered Window Structures," 2nd National Conference on Advances in Metrology (AdMet-2012), Pune, pp. 1-7, Feb. 2012.

Mahavir Singh, "Development of Lightweight Sandwich Material as Acoustic Partitions for Building Applications," Proc. of National Seminar on Material Characterization by Ultrasonics (NSMCU2012), New Delhi, pp. 1-10, Apr. 2012.

Dharam Pal Singh and Mahavir Singh, "Acoustic Properties of Coconut Coir Fiber Sound Absorptive Material," Proc. of National Seminar on Material Characterization by Ultrasonics (NSMCU2012), New Delhi, pp. 1-7, Apr. 2012. (Received a Best Paper Award).

Mahavir Singh and Dharam Pal Singh, "Evaluating Accurate Value of the Reverberation Chamber Sound Absorption Coefficient" Proc. of National Symposium on Acoustics (NSA2012), Tiruchengode, pp. 45-53, Dec. 2012

Mahavir Singh, "Acoustics Properties of New Rice Husk Sound Absorptive Material" Proc. of International Conference on Advances in Building Sciences" (BTCM 2013 Conferences), Chennai, pp. 257-264, Feb. 2013.

Mahavir Singh and Dharam Pal Singh, "Acoustic Metrology of Agricultural Waste Material as a Sound Absorber" Proc. of 8th International Conference on Advances in Metrology" (AdMet-2013), New Delhi, pp. 189-190, Feb. 2013.

Mahavir Singh, Kirti Soni and Gurbir Singh, "Noise Reduction Approach for the Compartment Ship" Proc. of International Conference on Acoustics (Acoustics 2013 New Delhi), New Delhi (in Press), Nov. 2013.

Mahavir Singh, Kirti Soni and Gurbir Singh, "Practical Acoustic Absorption Coefficient and Transmission Analysis of Multi-layer Absorbers" Proc. of International Conference on Acoustics (Acoustics 2013 New Delhi), New Delhi (in Press), Nov. 2013.

INDEX

ABOUT THE AUTHOR

Mahavir Singh was born in Ghaziabad (UP). He received his first degree in mechanical engineering, in 1988, from the National Institute of Technology (NIT), Surat. In 1991, he obtained an M. Tech. in mechanical engineering from the National Institute of Technology (NIT), Kurukshetra, and in 2003, obtained a Ph. D. degree in technical acoustics from the Indian Institute of Technology (IIT), Delhi. Currently he is a Principal Scientist in the Acoustics, Ultrasonics and Vibration (AUV) Section of the National Physical Laboratory, New Delhi and he has been involved in work of the acoustics and vibration fields for more than 20 years.

His main research interests are in the field of building acoustics (it is noise transmission in the audio frequencies), application of statistical energy analysis (SEA) to buildings and other structures, modelling of wall sound transmission losses using artificial neural network (ANN), noise and vibration control, calibration and testing of electro-acoustic equipments/ acoustical products and devices, maintenance, up-gradation and realization of national primary standards of acoustics to establish a quality system for traceable calibration measurement service; participation in key comparison of national acoustic standards at CCUVA/APMP level with NMIs of the world and advisory technical consultancy in architectural acoustics to provide services to the industry and institution of the Country. He has published more than 100 research papers in technical journals and conference proceedings. He is fellow of the Acoustical Society of India, has been guest editor for several journals and a member of several acoustical societies.